T0332880

Polynomial One-cocycles
for Knots and Closed Braids

Series on Knots and Everything — Vol. 64

Polynomial One-cocycles
for Knots and Closed Braids

Thomas Fiedler
Université Paul Sabatier, France

World Scientific

EW JERSEY · LONDON · SINGAPORE · BEIJING · SHANGHAI · HONG KONG · TAIPEI · CHENNAI · TOKYO

Published by

World Scientific Publishing Co. Pte. Ltd.
5 Toh Tuck Link, Singapore 596224
USA office: 27 Warren Street, Suite 401-402, Hackensack, NJ 07601
UK office: 57 Shelton Street, Covent Garden, London WC2H 9HE

British Library Cataloguing-in-Publication Data
A catalogue record for this book is available from the British Library.

Series on Knots and Everything — Vol. 64
POLYNOMIAL ONE-COCYCLES FOR KNOTS AND CLOSED BRAIDS

Copyright © 2020 by World Scientific Publishing Co. Pte. Ltd.

All rights reserved. This book, or parts thereof, may not be reproduced in any form or by any means, electronic or mechanical, including photocopying, recording or any information storage and retrieval system now known or to be invented, without written permission from the publisher.

For photocopying of material in this volume, please pay a copying fee through the Copyright Clearance Center, Inc., 222 Rosewood Drive, Danvers, MA 01923, USA. In this case permission to photocopy is not required from the publisher.

ISBN 978-981-121-029-7

For any available supplementary material, please visit
https://www.worldscientific.com/worldscibooks/10.1142/11551#t=suppl

Printed in Singapore

to Séverine,
Delphine,
Etienne

Was lange währt wird endlich gut
(old-fashioned saying)

Preface

Knots are as old as mankind. The first mathematical paper which has mentioned knots is "Remarques sur les problèmes de situation" by A.T. Vandermonde in 1771. Since then, mathematicians try to solve the main problem in knot theory: for two given smooth knots in 3-space (often called classical knots), decide in an effective way whether or not they are smoothly isotopic, i.e. can they be deformed one into the other in a differentiable way? The main tool to achieve this are knot invariants. There is an enormous number of known knot invariants, but either they are not calculable in an effective manner, i.e. they do not reduce to numbers which can be calculated by an algorithm in polynomial time, or they evidently fail to distinguish all knots. Hence the problem is still very far from its solution. Moreover, very little is known for the next question: given two closed 1-parameter families of knot diagrams (for long knots or knots in the solid torus), are they homotopic in the space of knot diagrams?

There were three main advances in 3-dimensional knot theory in the last decades: the discovery of quantum knot polynomials (and their upgradings to Topological Quantum Field Theories and categorifications), which come from representations of Lie algebras and more generally quantum groups; the discovery of Heegaard Floer homologies, which use deep results in complex analysis, and the discovery of finite type invariants (and the Kontsevich integral as their universal expression), which come from singularity theory. The singularity theory under question was Vassiliev's study of the discriminant of all singular smooth knots in the moduli space of all long knots in 3-space. But the approach from singularity theory disappeared again rapidly because of Dror Bar-Natans paper "On the Vassiliev knot invariants" which has transformed the subject back into representation theory of Lie algebras. The present book can be seen as a partial rehabilitation of singularity theory in knot theory.

ix

There is a natural projection of the 3-space into the plan and knots in 3-space can be given by knot diagrams, i.e. a smoothly embedded oriented circle in 3-space together with its projection into the plan. It is well known that oriented knot types are in one-one correspondence with knot types of oriented long knots. We close now the long knot with the 1-braid to a knot in the standardly embedded solid torus $V \subset \mathbb{R}^3$. Moreover, a framed long knot can be replaced by its parallel n-cable (with the same orientation of all strands) which we close to a knot in V by a cyclic permutation braid or more generally by a n-component string link T, which induces a cyclic permutation of its end points. The projection into the plan becomes now a projection pr into the annulus. We chose as generator of $H_1(V)$ the class which is represented by the closure of the oriented 1-braid. Hence the knot in V, which is obtained from the n-cable, represents the homology class $n \in H_1(V)$. We consider the infinite dimensional space M_n of all diagrams of knots K in V, which represent the homology class n, and such that there is a compressing disc of V which intersects K transversely and in exactly n points. M_n is called the *moduli space of knot diagrams without negative loops in the solid torus*. The knot types in 3-space correspond to the connected components of M_1. Given a generic knot diagram $K \subset V$ we consider the oriented curve $pr(K)$ in the annulus. A *loop in* $pr(K)$ is a piecewice smoothly oriented immersed circle in $pr(K)$ which respects the orientation of $pr(K)$. In other words, we go along $pr(K)$ following its orientation and at a double point we are allowed to switch perhaps to the other branch, but still following the orientation of $pr(K)$. Naturally, a loop in $pr(K)$ is called *negative* (respectively *positive*) if it represents a negative (respectively positive) homology class in $H_1(V)$. One easily sees that $pr(K)$ contains only positive loops if and only if $K \subset V$ is isotopic to a closed braid with respect to the disc fibration of V, and that knots which arise as cables of long knots by the above construction, contain never negative loops. The space M_n is a natural subspace of the space M of *all knots in the solid torus* and which represent the homology class n (without loss of generality we can assume that $n \geq 0$), i.e. negative loops in diagrams are now allowed as well.

An expression which is associated to a generic knot diagram in 3-space or in the solid torus is a knot invariant if and only if it is invariant under the Reidemeister moves. Many knot invariants were obtained in this way. We could interpret them as 0-cocycles on moduli spaces of diagrams. For example, the Alexander polynomial and the Jones polynomial are polynomial 0-cocycles on the moduli space of diagrams in 3-space. Goussarov has proven that all finite type invariants are integer 0-cocycles on the moduli space of diagrams of long knots (usually called Gauss diagram formulas). Let us call all such cocycles the *diagrammatic cocycles*.

It seems to us that the moduli spaces M_n and M have not been much studied before, but they turn out to be very important objects. The main goal of this book is to introduce a method which uses M_n and M in order to get new invariants for classical knots, for closed braids as well as for loops (up to homotopy) of knots and closed braids from diagrammatic 1-cocycles.

We have started to study the space M_n in 2006. The goal was to construct diagrammatic 1-cocycles instead of diagrammatic 0-cocycles. (It is well known that e.g. for a long hyperbolic knot K its component of M_1 deformation retracts onto a 2-dimensional torus, generated by two well-known loops. Consequently, cocycles of higher dimension would not contain more information in this case.) A diagrammatic 1-cocycle associates an integer to each Reidemeister move and the value of the 1-cocycle on a loop is the sum of the integers over all Reidemeister moves in the loop. A *quantized 1-cocycle* associates a monomial to each Reidemeister move instead of an integer. We have succeeded to define quantized 1-cocycles for closed braids but we have not succeeded in the case of knots in the solid torus (besides with an additional restriction on homotopies of isotopies). However, there are integer 1-cocycles in M_n and in M for $n \neq 1$ which depend non-trivially on a parameter $a \in (\mathbb{Z}/n\mathbb{Z}) \setminus \bar{0}$. We obtain then polynomial valued 1-cocycles simply by taking the Lagrange interpolation polynomial with respect to the parameter a.

The construction of the 1-cocycles is based on singularity theory together with rather complicated combinatorics. The singularity theory is the study of the discriminant Σ of all singular (i.e. non-generic) projections of a (non-singular) smooth oriented knot type in V into the annulus. (This discriminant is very different from Vassiliev's discriminant of singular knots.) In order to get knot invariants we apply the 1-cocycles to canonical loops in M_n respectively M. For example, for each knot in M there is the loop which is induced by the rotation of V around its core and for classical knots there is the loop in M_n which is induced by pushing the string link T once along the untwisted n-cable of the long knot.

It seems to us, that 3-dimensional knot theory splits into *two very different cases*: knots in orientable 3-manifolds for which non-trivial diagrammatic 1-cocycles exist (rather rare) and all the others (in some sense the generic case). Luckily, classical knots belong to the first case, because we can associate to them points in M_n. It could well be, that in the second case the usual finite type invariants are all what is needed in order to distinguish the objects. An example are braids. It is well known from results of Kohno and Bar-Natan, that finite type invariants indeed separate all braids. In this case the topological moduli space is essentially contractible (the only non-trivial loops consist of pushing full twists through other full-twists in a braid) and hence there are no non-trivial diagrammatic 1-cocycles for braids.

This book deals with the first of the two cases mentioned above and we concentrate on the construction of the diagrammatic 1-cocycles. We have tried to give their definitions as clear as possible, so that the reader can use them easily. The proofs of the invariance are rather direct but long. Therefore they are separated in sections for motivated readers. Calculations of examples by hand are rather tedious and we give only the minimal number of examples, but just enough to show that most of the 1-cocycles are not always trivial and that at least some of them are independent of each other. It is a very challenging problem to write computer programs to calculate the resulting invariants for many knots and to see what can be done with them. (Unfortunately, I'm not able to do this.)

The book is divided into an Introduction, which explains our method and puts it into a larger context, and three parts with the concrete results. The first part contains 1-cocycles for long knots, or more generally in M_n. The second part contains a 1-cocycle for all knots in the solid torus with $n \neq 1$. The third part contains quantized 1-cocycles in the special case of knots which are closed braids in the solid torus. We have kept the third part essentially self-contained, so that people how are more interested in braids than in knots can restrict themselves to the third part alone.

Before this book there were known explicitly only two non-trivial 1-cocycles for knots, namely the Teiblum-Turchin 1-cocycle for long knots and Budney's 1-cocycle for long knots which is based on Gramain's loop. The Teiblum-Turchin 1-cocycle is an integer 1-cocycle of finite type of degree 3. Vassiliev represented its reduction mod 2 as a diagrammatic 1-cocycle, Sakai represented the 1-cocycle over the reals by configuration space integrals and Mortier represented it as an integer diagrammatic 1-cocycle. Budney's integer 1-cocycle is not a diagrammatic 1-cocycle.

I wish to thank Ryan Budney and Victor Turchin for patiently explaining to me the topology of moduli spaces of knots, Hugh Morton for his observation that the graph in what has become the cube equations is indeed the 1-skeleton of a cube, my former co-author Vitaliy Kurlin for his deep immersion into singularity theory, Stepan Orevkov for his ideas about Garside's theory for braids and Alexander Stoimenow for his computer program in the case of closed braids. I am grateful to Dror Bar-Natan for first observing a "telescoping effect", which implies that many of my previous 1-cocycles for long knots were actually 1-coboundaries. Finally, let me mention that without Séverine, who has encouraged me over all these years and who has in particular created most of the figures, this book wouldn't exist.

Thomas Fiedler
Toulouse, June 2019

Contents

List of Figures

1

Introduction

The invariants in this book have their origin in a slight change of the subject: we study 1-parameter families of knots instead of individual knots. This is not an unusual feature in mathematics: for a difficult problem it is sometimes useful to attack an even more difficult problem (e.g. finding invariants for loops in the moduli space of all diagrams of knots). The topological moduli space of a knot in an oriented 3-manifold is the infinite dimensional space of all smooth knots smoothly isotopic to the given knot. If the 3-manifold is a line bundle over a surface then we consider the projection pr along the lines onto the surface. In this book we will assume that the line bundle is trivial and hence the surface is orientable (e.g. $pr : \mathbb{R}P^3 \setminus \{pt\} \to \mathbb{R}P^2$ stays an interesting case for further investigations). Moreover, we fix an orientation of the lines and hence of the surface. Each point in the moduli space is now a knot diagram with respect to pr and we call the moduli space with this additional structure the *moduli space of knot diagrams*, denoted by M. The connected component of M which corresponds to a knot type K is sometimes denoted by M_K. The projection pr induces a stratification on M. The strata of codimension 0 are the ordinary knot diagrams and their complement in M is the discriminant of singular knot diagrams, denoted by Σ. An oriented generic loop in M intersects Σ transversely in a finite number of points, which belong all to strata $\Sigma^{(1)}$ of codimension 1 in M. These strata correspond exactly to the Reidemeister moves, see e.g. [12]. All our diagrammatic 1-cocycles are of the following form: we associate an integer to each point in $s \cap \Sigma^{(1)}$, i.e. the Reidemeister moves, and we associate nothing to the ordinary diagrams.

Previously, our starting point was the Kauffman bracket for the Jones polynomial, see e.g. [35]. It is so simple and so beautiful, that the idea was to replace in its construction ordinary crossings by triple crossings. They allow much more different smoothings than ordinary crossings, which could be used to define state sums. But triple crossings are only stable in 1-parameter families. Therefore we have to construct 1-cocycles instead of 0-cocycles. However, the result was essentially only a 1-coboundary, see [18].

There can be diagrammatic 1-cocycles which represent a non-trivial cohomology class in M only under two conditions:

- *(a) $H_1(M)$ contains elements of infinite order*
- *(b) Let $c \in H_1(M)$ be a class of infinite order. Then each generic loop s which represents c intersects $\Sigma^{(1)}$.*

The conditions (a) and (b) are satisfied in two important cases:

- non-trivial long knots in 3-space, or what is the same, knotted 1-tangles in the 3-ball
- knots in the solid torus which are not contained in a 3-ball and which are not isotopic to the core of the solid torus.

In this book we construct diagrammatic 1-cocycles for all knots in these two cases.

There is at least one other case which satisfies (a): let the K be a knot in a thickened torus (e.g. the regular neighborhood of an incompressible torus of the JSJ-decomposition of an orientable 3-manifold) and which is not contained in a 3-ball nor in a solid torus other than its regular neighborhood. There are loops of infinite order in M_K induced by the rotations of the torus in itself in the thickened torus, see e.g. [31]. But these loops consist only of isotopic diagrams (for the natural projection pr into the torus) and hence there are no diagrammatic 1-cocycles in this case (these loops do not intersect the discriminant $\Sigma^{(1)}$ at all, i.e. (b) is not satisfied).

Let M^3 be a closed connected orientable 3-manifold. It follows easily from the prime decomposition theorem of Kneser and Milnor for orientable 3-manifolds [39] together with the theorem of Gordon and Luecke that unoriented knots in S^3 are determined by their complements [24], that two unoriented local knots (i.e. contained in a 3-ball) are ambient isotopic in M^3 if and only if they are isotopic as local knots. Hence, we can apply our invariants to them (more precisely, to the couple of the oriented knot and its inverse).

Let K be a knot in M^3 which is contained in a solid torus V other than its regular neighborhood. The entrance door for our invariants to knots in general 3-manifolds would be that the following conjecture holds.

Conjecture 1.1 *Let V be a solid torus in M^3. We assume that the core of V does not represent an element of order at most 2 in the fundamental group of M^3 and that ∂V is incompressible from the outside, i.e. in $M^3 \setminus Int(V)$. Let K and K' be knots in V which are isotopic in M^3 and which are not contained in a 3-ball. Then K and K' are already isotopic in V.*

(An example of Motegi [44] shows that the condition about the non-triviality of the core in the fundamental group is necessary. If the core does not represent an element of order 2 then it can not be an invertible knot in M^3 (i.e. it is not ambient isotopic to itself with reversed orientation), what is evidently a necessary condition. Certain homotopically non-trivial knots in $S^1 \times S^2$ and the light bulb trick show that the incompressibility condition is necessary too.)

If true, the conjecture would imply that isotopy invariants of K in V are isotopy invariants of K in M^3. Hence, we could apply our invariants for knots in the solid torus to $K \subset M^3$ as well.

It is amazing that in contrast to quantum knot invariants our invariants do not apply to virtual knots (because of the forbidden moves in each loop, see [36]) neither to generic unordered links. Indeed, let $K \cup K'$ be a sufficiently general 2-component link in the 3-sphere. As well known, links are not equivalent to long links. On the other hand, we can consider K as a knot in the 3-manifold $S^3 \setminus K'$. But in the "generic case" the moduli space of K in $S^3 \setminus K'$ will be contractible and hence the above condition (a) is not satisfied.

The topology of moduli spaces of classical knots and of long knots was much studied in [29], [8], [9], [10], [11]. In particular, there is a complete description of the homotopy type and of the homology groups (even with an additional structure of certain Gerstenhaber-Poisson algebras) for the space of long knots in $\mathbb{R}^3$ (the same space, which was studied by Vassiliev using singularity theory) in the subsequent papers [29], [9], [10], [11]. Vassiliev [54] had constructed certain universally defined 0-dimensional cohomology classes. They give the finite type invariants of long knots (or equivalently of compact knots in the 3-sphere) when they are evaluated on the 0-dimensional homology classes of the disconnected space. It is known that all finite type knot invariants, also called Vassiliev-Goussarov invariants, have diagrammatic formulas, which allow the computation of the invariant from an arbitrary generic knot diagram [27]. Our goal is the construction of diagrammatic formulas for 1-dimensional cohomology classes of the space of long knots, which give knot invariants when they are applied to loops, represented by generic 1-parameter families of diagrams for long knots. It seems that the Teiblum-Turchin 1-cocycle v_3^1 was the only known 1-cocycle for long knots which represents a non-trivial cohomology class and which has (probably) an explicit diagrammatic formula for its computation. The Teiblum-Turchin 1-cocycle is an integer valued 1-cocycle of degree 3 in the sense of Vassiliev's theory [54]. Its reduction mod 2 had the first diagrammatic description of a 1-cocycle, see [55] and [53]. Sakai has defined a $\mathbb{R}$ valued version of the Teiblum-Turchin 1-cocycle via configuration space integrals [49]. We have found a very complicated formula for an integer extension of v_3^1 mod 2 in [18]. The most beautiful diagrammatic formula for an integer valued 1-cocycle for long knots which extends v_3^1 mod 2 and which probably coincides with v_3^1 was found by Mortier [40], [41]. There is also a generalization of the Kontsevich integral to a 1-cocycle by Mortier [42]. Budney has defined another integer valued 1-cocycle on the space of long knots in [10], Propositions 6.1 and 6.3, using Gramain's loop. The value of his 1-cocycle on Gramain's loop (see below for a definition) for a knot K is n if K is the connected sum of exactly n non-trivial knots. But it is not clear how to represent it by a diagrammatic 1-cocycle.

Chapter 2 deals with classical knots. It is well known that if a knot K in the 3-sphere is not a satellite then its topological moduli space is a $K(\pi, 1)$ with π a finite group (in fact $\pi = Aut(\pi_1(S^3 \setminus K), \partial)$, where ∂ is the peripheral system of the knot K, see [56], [31], [30]). Consequently, in this case there can't exist any non-trivial 1-cocycles with values in a torsion free module. However, it is also well known that the components of the moduli space of knots in the 3-sphere are in a natural $1 - 1$ correspondence with the components of the moduli space of long knots in 3-space. Hatcher [29] has proven that for long knots in 3-space the situation is much better. There are always two canonical non-trivial loops in the component of the topological moduli space of a long knot K if it is not the trivial knot: Gramain's loop, denoted by $rot(K)$, and the Fox-Hatcher loop, denoted by $fh(K)$. *Gramain's loop* is induced by the rotation of the 3-space around the long axis of the long knot [26]. *Fox-Hatcher's loop* is defined as follows: one puts a pearl (i.e. a small 3-ball B) on the closure of the long framed knot K in the 3-sphere. The part of K in $S^3 \setminus B$ is a long knot. Pushing B once along the knot with respecting the framing induces the Fox-Hatcher loop, see [29] and also [22]. The homology class of rot(K) does not depend on the framing of K and changing the framing of K adds multiples of rot(K) to fh(K). Notice that the Fox-Hatcher loop has a canonical orientation induced by the orientation of the long knot. The same loops are still well defined and non-trivial for those n-string links which are n-cables of a framed non-trivial long knot. It follows from results of Hatcher [29] and Budney [11] that these two loops are linearly dependent in the rational homology if and only if the knot is a torus knot (compare also Lemma 2.1 in Section 2.3 and Proposition 4.9 in Section 4.8.3). Moreover, Hatcher has shown that the topological moduli space of a long hyperbolic knot deformation retracts onto a 2-dimensional torus. Hence it follows from Künneth's formula that it is sufficient to construct just 1-cocycles in this case.

Let us try to make it completely clear to the reader: the homology of the space of long knots depends very strongly on the component of the space. But luckily, each component contains two canonical 1-homology classes, namely the class represented by Gramain's loop and the class represented by the Fox-Hatcher loop (which are dependent or even trivial only in very particular cases). We can represent these loops and their natural generalizations in M_n and in M by 1-parameter families of diagrams. We are now interested in those 1-cocycles in M_n and in M which are universally defined (i.e. independent of the component of the space) and which can be calculated from the 1-parameter family of diagrams alone.

Definition 1.1 *We fix a orthogonal projection $pr : \mathbb{C} \times \mathbb{R} \to \mathbb{C}$ together with standard coordinates (x, y, z) of $\mathbb{C} \times \mathbb{R}$. A* long knot K *is an oriented smoothly embedded copy of $\mathbb{R}$ in $\mathbb{C} \times \mathbb{R}$ which coincides with the real x-axis in $\mathbb{C} \times 0$ outside a compact set. A* parallel n-cable *of a framed long knot is a n-component link with fixed endpoints where each component is parallel to the framed long knot with respect to the blackboard framing given by pr (the z-coordinate) and with the same orientation on each component. A* n-string link T *is a n-component link with fixed endpoints where each component is parallel to a long knot in $\mathbb{C} \times 0$ outside some compact set.*

We cut now the string link T with a very big 3-ball B_1. The end points of the string link are respectively in two big discs and we glue another 3-ball B_2 to the discs in order to obtain a solid torus V. We chose a string link $S \subset B_2$ such that $T \cup S$ is an oriented knot $K \subset V$.

Definition 1.2 *We denote by M is the topological moduli space of all oriented smooth knots in V which represent the homology class $n \in H_1(V)$.*

Each point in M comes with its projection pr, i.e. is represented by a knot diagram. *We denote by M_n be the subspace of all knots without negative loops in the knot diagram and by M_n^+ the subspace of all knots for which each loop in the diagram represents a strictly positive homology class.*

Evidently, $M_n^+ \subset M_n \subset M$, and the inclusions are strict. It turns out that M_n and also M are the right spaces in order to study classical knots, and that M_n^+ is the right space in order to study closed braids in the solid torus. If the string link T is the parallel n-cable of a long framed knot k then evidently $K = T \cup S$ does not contain any negative loops. Moreover, if a framed long knot k' is (framed) isotopic to k then the corresponding knots $K = T \cup S$ and $K' = T' \cup S$ are in the same component of M_n. Consequently, each 1-cocycle of M_n evaluated on an universally defined loop in M_n (i.e. the definition of the loop does not depend on the component of M_n) gives an invariant of classical knots. If we change the orientation of some of the components of the parallel n-cable then the resulting knot $K = T \cup S$ is no longer in M_n but in M and again each 1-cocycle of M evaluated on an universally defined loop gives an invariant of classical knots.

But of course, the 1-cocycles can distinguish also loops. If a 1-cocycle has different values on two loops then these loops are not homotopic (and even not homologous, because the values are in an abelian ring) in M_n respectively in M.

It is an interesting open question whether the inclusions $in : M_n^+ \subset M_n \subset M$ are injective for H_0 and H_1. It seems that it is only known that the inclusion is injective for $H_0(M_n^+) \to H_0(M)$, i.e. if we restrict ourselves only on closed n-braids. This means that two closed braids are isotopic as knots in the solid torus if and only if they are isotopic as closed braids (and hence represented by the closure of conjugate braids), see e.g. [43].

It is often convenient for calculations to represent a long knot K as a closed braid with just one strand opened to go to infinity.

In this book we construct the first non-trivial integer diagrammatic 1-cocycles in M_n and in M, and which depend on an integer parameter. For each fixed n they can be calculated with at most $O(c^3)$ operations with respect to the maximal number of ordinary crossings c of a diagram among all diagrams in the 1-parameter family. We construct also the first polynomial diagrammatic 1-cocycles in M_n^+ which can distinguish a closed braid from its inverse (what quantum invariants fail to do).

It follows from Thom-Mather singularity theory that each component of the infinite dimensional space M has a natural *stratification with respect to* pr:

$$M = \Sigma^{(0)} \cup \Sigma^{(1)} \cup \Sigma^{(2)} \cup \Sigma^{(3)} \cup \Sigma^{(4)}...$$

Here, $\Sigma^{(i)}$ denotes the union of all strata of codimension i.

The strata of codimension 0 correspond to the usual generic *diagrams of knots*, i.e. all singularities in the projection are ordinary double points. So, our *discriminant* is the complement of $\Sigma^{(0)}$ in M. Notice that this discriminant of non-generic diagrams is very different from Vassiliev's discriminant of singular knots [54].

The three types of strata of codimension 1 correspond to the *Reidemeister moves*, i.e. non generic diagrams which have exactly one ordinary triple point, denoted by $\Sigma_{tri}^{(1)}$, or one ordinary self-tangency, denoted by $\Sigma_{tan}^{(1)}$, or one ordinary cusp, denoted by $\Sigma_{cusp}^{(1)}$, in the projection pr. We call the triple point together with the under-over information (i.e. its embedded resolution) a *triple crossing*. We distinguish self-tangencies for which the orientation of the two tangents coincide, called $\Sigma_{tan+}^{(1)}$, from those for which the orientations of the tangents are opposite, called $\Sigma_{tan-}^{(1)}$.

Proposition 1.1 *There are exactly six types of strata of codimension 2. They correspond to non generic diagrams which have exactly either*

- *(1) one ordinary quadruple point, denoted by $\Sigma_{quad}^{(2)}$*

- *(2) one ordinary self-tangency with a transverse branch passing through the tangent point, denoted by $\Sigma_{trans-self}^{(2)}$*

- *(3) one ordinary self-tangency in an ordinary flex ($x = y^3$), denoted by $\Sigma_{self-flex}^{(2)}$*

- *(4) two singularities of codimension 1 in disjoint small discs (this corresponds to the transverse intersection of two strata from $\Sigma^{(1)}$, i.e. two simultaneous Reidemeister moves at different places of the diagram)*

- *(5) one ordinary cusp ($x^2 = y^3$) with a transverse branch passing through the cusp, denoted by $\Sigma_{trans-cusp}^{(2)}$*

- *(6) one degenerate cusp, locally given by $x^2 = y^5$, denoted by $\Sigma_{cusp-deg}^{(2)}$ We show these strata in Fig. 1.1.*

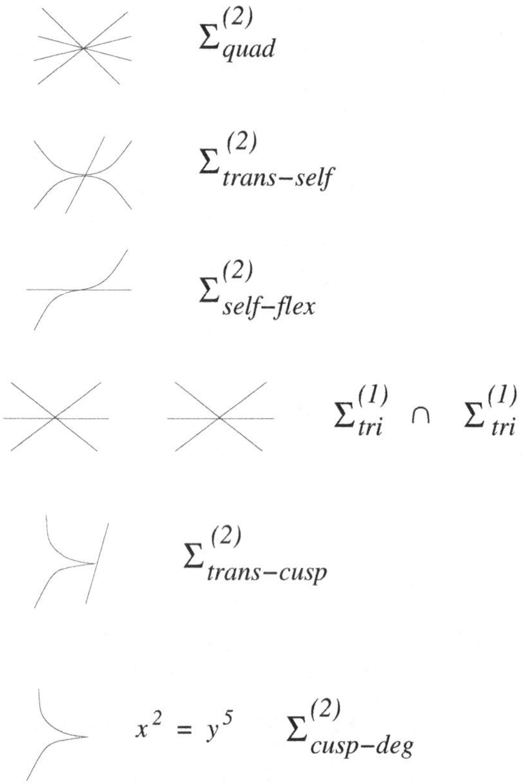

$$\Sigma^{(2)}_{quad}$$

$$\Sigma^{(2)}_{trans-self}$$

$$\Sigma^{(2)}_{self-flex}$$

$$\Sigma^{(1)}_{tri} \cap \Sigma^{(1)}_{tri}$$

$$\Sigma^{(2)}_{trans-cusp}$$

$$x^2 = y^5 \quad \Sigma^{(2)}_{cusp-deg}$$

Figure 1.1: The strata of codimension 2 of the discriminant of non generic projections

The basic notions of our one parameter approach to knot theory, namely the space of non-singular knots, its discriminant, the stratification of the discriminant, the unfoldings of the strata in terms of singularity theory, the coorientation of strata of low codimension, the canonical loop, the trace graph, the equivalence relation for trace graphs, were worked out in all details in our joint work with Vitaliy Kurlin [20], [21] (compare also [13]). For completeness and for the convenience of the reader we have add this work in Sections 4.1 up to 4.6.

Our strategy is the following: for an oriented generic loop in M we associate an integer to the intersection with each stratum in $\Sigma^{(1)}_{tri}$, i.e. to each Reidemeister III move, and we sum up over all moves in the loop.

We obtain three sorts of 1-cocycles in the case of classical knots: the first one does not depend on any parameter and is in fact a generalization of the Mortier and Teiblum-Turchin 1-cocycle for long knots. The second one depends non-trivially on an integer parameter $0 < a < n$. The third one depends on two integer parameters a and b with $0 < a < n$, $0 < b < n$ and $a+b > n$. However, we do not know whether this 1-cocycle is non-trivial too.

In order to show that our 1-cochains are 1-cocycles, we have to prove that the sum is 0 for each meridian of strata in $\Sigma^{(2)}$. This is very complex but we use strata from $\Sigma^{(3)}$ in order to reduce the proof to a few strata in $\Sigma^{(2)}$. It follows that our sum is invariant under generic homotopies of loops in M. But it takes its values in an abelian ring and hence it is a 1-cocycle. Showing that the 1-cocycle is 0 on the meridians of $\Sigma^{(2)}_{quad}$ is by far the hardest part. This corresponds to finding a new solution of the *tetrahedron equation*. Consider four oriented straight lines which form a braid and such that the intersection of their projection into $\mathbb{C}$ consists of a single point. We call this an *ordinary quadruple crossing*. After a generic perturbation of the four lines we will see now exactly six ordinary crossings. We assume that all six crossings are positive and we call the corresponding quadruple crossing a *positive quadruple crossing*. Quadruple crossings form smooth strata of codimension 2 in the topological moduli space of lines in 3-space which is equipped with a fixed projection pr. Each generic point in such a stratum is adjacent to exactly eight smooth strata of codimension 1. Each of them corresponds to configurations of lines which have exactly one ordinary triple crossing besides the remaining ordinary crossings. We number the lines from 1 to 4 from the lowest to the highest (with respect to the projection pr). The eight strata of triple crossings glue pairwise together to form four smooth strata which intersect pairwise transversely in the stratum of the quadruple crossing, see e.g. [20]. The strata of triple crossings are determined by the names of the three lines which give the triple crossing. For shorter writing we give them names from P_1 to P_4 and $\bar{P}_1$ to $\bar{P}_4$ for the corresponding stratum on the other side of the quadruple crossing. We show the intersection of a normal 2-disc of the stratum of codimension 2 of a positive quadruple crossing with the strata of codimension 1 in Fig. 1.2. The strata of codimension 1 have a natural coorientation, compare the next section. We could interpret the six ordinary crossings as the edges of a tetrahedron and the four triple

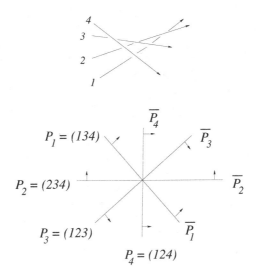

Figure 1.2: The intersection of a normal 2-disc of a positive quadruple crossing with the strata of triple crossings

crossings likewise as the vertices or the 2-faces of the tetrahedron. For the classical tetrahedron equation one associates to each stratum P_i, i.e. to each vertex or equivalently to each 2-face of the tetrahedron, some operator (or some R-matrix) which depends only on the names of the three lines and to each stratum $\bar{P}_i$ the inverse operator. The tetrahedron equation says now that if we go along the meridian then the composition of these operators is equal to the identity. Notice, that in the literature, see e.g. [34], one considers planar configurations of lines. But this is of course equivalent to our situation because all crossings are positive and hence the lift of the lines into 3-space is determined by the planar picture. Moreover, each move of the lines in the plane which preserves the transversality lifts to an isotopy of the lines in 3-space. The tetrahedron equation has many solutions, the first one was found by Zamolodchikov, see e.g. [34].

However, the solutions of the classical tetrahedron equation are not well adapted in order to construct 1-cocycles for moduli spaces of knots. First of all there is no natural way to give names to the three branches of a triple crossing in an arbitrary knot isotopy besides in the case of closed braids. But it is not hard to see that in the case of braids Markov moves would make big trouble (see e.g. [5] for the definition of Markov moves and Markovs theorem). Secondly, a local solution of the tetrahedron equation is of no use for us, because as already pointed out there are no integer valued 1-cocycles for knots in the 3-sphere. We have to replace them by long knots or more generally by points in M_n. For knots in the solid torus V we can now associate to each crossing in the diagram a winding number (i.e. a homology class in $H_1(V)$) in a canonical way. Therefore we have to consider six different positive tetrahedron equations, corresponding to the six different abstract closures of the four lines to a circle and in each of the six cases we have to consider all possible winding numbers of the six crossings. We call this the *positive global tetrahedron equations*. One easily sees that there are exactly *forty eight* local types of quadruple crossings (analog to the eight local types of triple crossings). We study the relations of the local types in what we call the *cube equations*.

Chapter 3 deals with all knots in the solid torus V, i.e. points in M. It turns out that exactly one of our 1-cocycles in M_n can be generalized to a 1-cocycle in M. It is automatically trivial if the knot is isotopic to a knot contained in a 3-ball in V and it is also automatically trivial if the knot is isotopic to a closed braid in V. After identifying $S^1 \times I^2$ with $S^1 \times D^2$ there are two universally defined canonical loops in each M_K: the analog of Gramain's loop $rot(K)$ which is induced by the rotation of each disc $p \times D^2$ around its center and the loop $slide(K)$ which is induced by the rotation of the core in itself (i.e. we slide the knot along the core of the solid torus). Evidently, $slide(K)$ does not intersect $\Sigma^{(1)}$ at all (it is just an isotopy of diagrams) and hence our 1-cocycle is 0 on it.

Definition 1.3 *Let K be a knot in the solid torus V which is not contained in a 3-ball nor isotopic to the core of the solid torus. K is called* periodic *if $[rot(K)]$ is a non-trivial rational multiple of $[slide(K)]$ in $H_1(M_K; \mathbb{Q})$.*

It is well known that if the knot K is a closed braid in the solid torus (i.e. intersects all discs $p \times D^2$ transversely) then K is periodic in our sense if and only if the braid is periodic in the usual sense (i.e. there exists a power of it which is isotopic to some power of the full twist Δ^2, compare Chapter 4). Our 1-cocycle can sometimes be used to show very easily that a knot $K \subset V$ is not periodic (because it is always 0 on $slide(K)$).

Chapter 4 deals only with those knots which are closed braids in the solid torus, and it is essentially based on [19] and [20].

I had constructed integer valued finite type 1-cocycles for closed braids in [17]. In fact, these invariants have a completely evident "quantization" to polynomial valued invariants, which are no longer of finite type but still calculable with polynomial complexity [19]. They can detect sometimes very easily the non-invertibility of a closed braid, what quantum knot invariants fail to do, (that is the non-invertibility of the link in S^3 which consists of the closed braid together with the closure of the braid axes), when it is applied to the loop $rot(K)$ generated by the rotation of the solid torus around its core, compare [17] and [19]. An important point is that all our invariants can be calculated with polynomial complexity with respect to the number of crossings of the braids (i.e. the geometric braid length).

As well known, the isotopy problem for closed braids in the solid torus reduces to the conjugacy problem in braid groups (see e.g. [43]). The latter problem is solved, but in general only with exponential complexity with respect to the braid length (see [23] and [6]). It is therefore interesting to construct invariants which distinguish conjugacy classes of braids and which are calculable with polynomial complexity. Finite type invariants for knots in the solid torus are an example of such invariants (see [54], [55], [2], [25], [16] and references therein).

In Chapter 4 we construct two new classes of invariants for closed braids, which are calculable in polynomial time: *one-cocycle polynomials and character invariants*. Moreover, our new invariants are more related to geometric invariants of braids (as entropy and simplicial volume of its mapping torus) than are the usual invariants (e.g. the HOMFLYPT polynomial for links in the solid torus, compare [52] and [25] for its definition): if for an irreducible braid one of our invariants is non-trivial then the braid is pseudo-Anosov, and hence both its entropy and its hyperbolic volume are non-trivial (compare Proposition 4.9).

We give now a brief outline of our approach in this chapter, which is a quit different from the approach in the previous chapters.

Let $\hat{\beta} \subset V$ be a closed braid (i.e. it intersects each disc in the fibration $V = S^1 \times D^2$ transversely) and such that $\hat{\beta}$ is a knot. Let $M_n^+ \subset M_n$ be now the infinite dimensional space of all closed n-braids in V and which are knots (i.e. all loops in the diagrams are now strictly positive). First, we associate to a closed braid in a canonical way a loop in M_n^+, called $rot(\hat{\beta})$, namely the loop induced by the rotation of the solid torus around its core, still called Gramain's loop or the canonical loop. It has a very nice combinatorial description, namely it is just the pushing of a full-twist Δ^2 through the closed braid. We associate to Gramain's loop an oriented singular link in a thickened torus by tracking the crossings of the closed braids in the loop. This link is called the *trace graph* and it is denoted by $TL(rot(\hat{\beta}))$. All singularities of $TL(rot(\hat{\beta}))$ are ordinary triple points. These triple points correspond exactly to the intersections of $rot(\hat{\beta})$ with $\sum_{tri}^{(1)}$. There is a natural coorientation on $\sum_{tri}^{(1)}$ and, hence, each triple point in $TL(rot(\hat{\beta}))$ has a sign. To each triple point corresponds a diagram of a closed braid which has just ordinary crossings and exactly one triple crossing. We use the position of the ordinary crossings with respect to the triple crossing in the *Gauss diagram* in order to construct monomials as *weights* for the triple crossings.

We associate then to each generic loop $\gamma \subset M_n^+$ *polynomial invariants* $\Gamma(\gamma) \in \mathbb{Z}[x, x^{-1}]$ *which depend only on the homology class of* $\gamma \subset M_n^+$*, called one-cocycle polynomials, by summing over the weighted intersection numbers of* γ *with* $\sum_{tri}^{(1)}$. In particular, $\Gamma(\gamma) \in \mathbb{Z}[x, x^{-1}]$ depends only on the Reidemeister III moves in γ and does not depend on the Reidemeister II moves at all.

We show that in particular $\Gamma(\hat{\beta}) = \Gamma(rot(\hat{\beta}))$ is not of finite type but that $d(\Gamma(\hat{\beta}))/dx$ evaluated at $x = 1$ is a integer valued finite type invariant of $\hat{\beta}$. We call them *finite type one-cocycle invariants*. As all finite type invariants, the finite type one-cocycle invariants have a natural degree. However, our invariants induce a new filtration on the space of all finite type invariants for closed braids, as shows the following fact: Let $\hat{\beta}$ be a closed n-braid which is a knot and let c be the word length of $\beta \in B_n$ (with respect to the standard generators of B_n, or in other words, the number of crossings of the diagram with respect to pr). Then all finite type one-cocycle invariants of degree d vanish for $\hat{\beta}$ if $d \geq c + n^2 - n - 1$ (Proposition 4.5).

One has to compare this with the following well known fact: already the trefoil has non-trivial finite type invariants of arbitrary high degree. Consequently, the finite type one-cocycle invariants define a new natural filtered subspace in the filtered space of all finite type invariants for those knots in the solid torus, which are closed braids.

Our polynomial invariants for closed braids in the solid torus are based on extremely simple new solutions of the tetrahedron equation in M_n^+, without solving a big system of equations, in contrast to the case of knots in the solid torus. (The corresponding finite type invariants, namely their values

at $x = 1$, satisfy automatically the *marked 4T-relations*, compare [25] and
also [16].) We define a polynomial one-cocycle R in Section 4.8.5 which can
detect with quadratic complexity the non-invertibility of a closed braid in
the solid torus.

Moreover, it turns out that the one-cocycle invariants can be refined con-
siderably. When we deform $\hat{\beta}$ in V by a generic isotopy then $rot(\hat{\beta})$ in M_n^+
deforms by a generic homotopy. The following is our key observation for the
refinement (compare Remark 4.2).

Observation 1.1 *The loop $rot(\hat{\beta})$ is never tangential to $\sum_{tan}^{(1)}$.*

(In other words, in a homotopy of the loop $rot(\hat{\beta})$ we have never to undo
a Reidemeister II move just after we have done the move.) Let l be an integer
and let us consider l iterations of the loop $rot(\hat{\beta})$. It follows from this observation that the connected components of the natural resolution of $TL(l.rot(\hat{\beta}))$
(i.e. the abstract union of circles where the branches in the triple points are
separated) are isotopy invariants of $\hat{\beta} \hookrightarrow V$. We apply now our theory of
one-cocycle polynomials but only to those triple points in $TL(l.rot(\hat{\beta}))$ where
three *given* components (of the resolution) of $TL(l.rot(\hat{\beta}))$ intersect. The resulting invariants are called *character invariants* (Theorems 4.8 and 4.9).
Character invariants take into account the permutation (called *monodromy*)
of the crossings of $\hat{\beta}$ which is induced by the loop $l.rot(\hat{\beta})$. A character
invariant for a fixed l corresponds now to an unordered set of one-cocycle
polynomials, namely one polynomial for each fixed triple of components (not
necessarily different) of the resolution of $TL(l.rot(\hat{\beta}))$. The character invariants for $l = 1$ coincide with the one-cocycle polynomials, as follows from
Lemma 4.20. Taking l multiples of the loop $rot(\hat{\beta})$ changes the character
invariants in an uncontrollable way because of the monodromy of the crossings, which depends strongly on the closed braid $\hat{\beta}$. Hence, each character
invariant forms no longer a one-cocycle, but just a map $\mathbb{Z} \to \{$ *unordered sets
of integer Laurent polynomials*$\}$. These new invariants are not of finite type
but they are still calculable with polynomial complexity with respect to the
braid length.

Alexander Stoimenow has written a computer program in order to calculate the simplest character invariants (see Section 4.9.3). It turns out that
already these character invariants of linear complexity can sometimes detect
the non-invertibility of closed braids.

2

1-cocycles for classical knots

2.1 The 1-cocycles

The local types of Reidemeister moves for unoriented knots are shown in Fig. 2.1.

For oriented knots there are eight local types of R III moves (compare Fig. 2.31), four local types of R II moves and four local types of R I moves. Different local types of Reidemeister moves come together in strata of $\Sigma^{(2)}$. Their relations will be extensively studied in Sections 2.4.3 and 2.4.4.

To each Reidemeister move of type III corresponds a diagram with a *triple crossing* p: three branches of the knot (the highest, middle and lowest with respect to the projection $pr : \mathbb{C}^* \times \mathbb{R} \to \mathbb{C}^*$) have a common point in the projection into the plane. A small perturbation of the triple crossing leads to an ordinary diagram with three crossings near $pr(p)$.

Definition 2.1 *We call the crossing between the highest and the lowest branch of the triple crossing p the* distinguished crossing *of p and we denote it by d (d stands for distinguished). The crossing between the highest branch and the middle branch is denoted by hm and that of the middle branch with the lowest is denoted by ml, compare Fig. 2.2. For better visualization we draw the crossing d always with a thicker arrow.*

Smoothing an ordinary crossing c of a diagram D with respect to the orientation splits the closure of D into two oriented and ordered circles. We call D_c^+ the component which goes from the under-cross to the over-cross at c and by D_c^- the remaining component, compare Fig. 2.3.

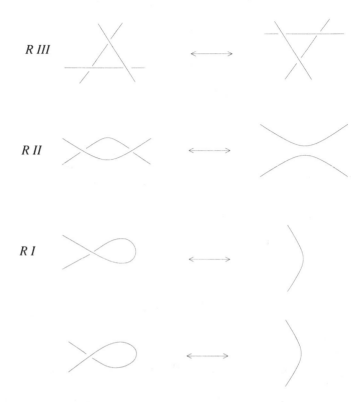

Figure 2.1: The Reidemeister moves for unoriented knots

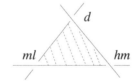

Figure 2.2: The names of the crossings in a R III-move

Figure 2.3: Two ordered knot diagrams associated to a crossing c

We associate to each generic crossing q in a diagram the integer $[q] = [D_q^+] \in \mathbb{Z} = H_1(V)$ for the above mentioned identification. The integer $[q]$ is called the *homological marking* of q.

A *Gauss diagram of* a knot $K \subset V$ is an oriented circle with oriented chords which are decorated with their homological markings. The chords correspond to the crossings of the knot diagram and are always oriented from the under cross to the over cross (here we use the orientation of the $\mathbb{R}$-factor). Moreover, each chord (or arrow) has a sign, which corresponds to the writhe (or sign) of the crossing. The circle of a Gauss diagram in the plane is always equipped with the counter-clockwise orientation.

A *Gauss diagram formula* of degree k is an expression assigned to the diagram of a knot $K \subset V$, which is of the following form:

$$\sum_{configurations} \text{function(writhes of the crossings)}$$

where the sum is taken over all possible choices of k (unordered) different crossings in the diagram such that the chords arising from these crossings in the diagram of K build a given sub-diagram with given markings, but without fixing the signs on the arrows. The marked sub-diagrams are called *configurations*, compare e.g. [47], [16], [19]. If the function is (as usual) the product of the writhes of the crossings in the configuration, then we will denote the sum shortly by the configuration itself. As usual, the *writhe* (or handedness, or sign) of a positive crossing (see e.g. Fig. 2.3) is $w = +1$ and the writhe of a negative crossing is $w = -1$.

A Reidemeister III move corresponds to a triangle in the Gauss diagram. The *global type of a Reidemeister III move* is now shown in Fig. 2.4, where $m+h$, $m+h-[K]$, m and h are the homological markings of the corresponding arrows. Here, $[K]$ is the homology class represented by the knot K. Using the homological markings for a given $n \neq 0$, we denote the corresponding strata in $\Sigma_{tri}^{(1)}$ simply by $\Sigma_{([d],[hm],[ml])}^{(1)}$. Only for $n = 0$ we have in addition to indicate whether the arrow ml in the triangle goes to the left, denoted by l, or it goes to the right, denoted by r. We will consider Gauss diagram formulas for R III moves. The arrows in the configurations which are not arrows of the triangle are called the *weights* of the formulas.

Definition 2.2 *The* coorientation *for a Reidemeister III move is the direction from two intersection points of the corresponding three arrows to one*

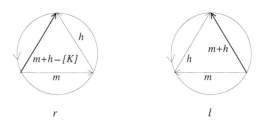

Figure 2.4: The global types of R III moves for knots in the solid torus

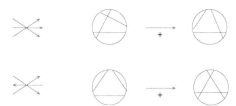

Figure 2.5: The coorientation for Reidemeister III-moves

*intersection point and of no intersection point of the three arrows to three in-
tersection points, compare Fig. 2.5. (We will see later in the cube equations
for $\Sigma^{(2)}_{trans-self}$ that the two coorientations for triple crossings fit together for
the strata of $\Sigma^{(1)}_{tri}$ which come together in $\Sigma^{(2)}_{trans-self}$.) Evidently, our coorientation is completely determined by the corresponding planar curves and
therefore we can draw just chords instead of arrows in Fig. 2.5. We call the
side of the complement of $\Sigma^{(1)}$ in M into which points the coorientation, the
positive side of $\Sigma^{(1)}$.*

*The coorientation for Reidemeister II and Reidemeister I moves is the
direction from no crossings to the diagram with two respectively one new
crossing.*

Each transverse intersection point p of an oriented generic arc in M with
$\Sigma^{(1)}_{tri}$ has now an intersection index $+1$ or -1, called $sign(p)$, by comparing
the orientation of the arc with the coorientation of $\Sigma^{(1)}$.

We are now ready to define our first 1-cocycles.

Let us consider the Newton 2-simplex in $\mathbb{R}^2$ which is shown in Fig. 2.6, i.e.
we consider all integer points (a, b) in the 2-simplex. We define a 1-cochain
R (the letter "R" stands for Reidemeister) for each open sub-simplex of the
2-simplex.

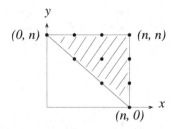

Figure 2.6: Newton polygon for our 1-cocycles

Let $\gamma \subset M_n$ be any oriented generic loop.

Definition 2.3 *(The 1-cochains $R_{(a,b)}$)*

Let $n > 0$.

The 1-cochain $R_{(a,b)}$ for the open 2-simplex $0 < a < n$, $0 < b < n$, $a + b > n$ is defined in Fig. 2.8.

The 1-cochain $R_{(a,n-a)}$ for the open 1-simplex $0 < a < n$, $a + b = n$ is defined in Fig. 2.9.

The 1-cochain $R_{(a,n)}$ for the open 1-simplex $0 < a < n$ is defined in Fig. 2.7.

The 1-cochain $R_{(n,b)}$ for the open 1-simplex $0 < b < n$ is defined in Fig. 2.10.

The 1-cochain $R_{(0,n)}$ for the 0-simplex $(0,n)$ is defined in Fig. 2.11.

The 1-cochain $R_{(n,0)}$ for the 0-simplex $(n,0)$ is defined in Fig. 2.12.

The 1-cochain $R_{(n,n)}$ for the 0-simplex (n,n) is defined in Fig. 2.13.

Our notation conventions are explained in Fig. 2.7.

Notice that the signs of the crossings in the triple crossing (i.e. in the triangle) do not enter the formulas. Our 1-cochains are defined in M_n, i.e. the space of all diagrams of oriented knots in the solid torus, which represent the homology class $n > 0$, and which do not contain loops in the diagram which represent a negative homology class (compare the Preface). As already explained, this space is well adapted in order to study classical knots.

Theorem 2.1 *All 1-cochains from Definition 2.3 are 1-cocycles in M_n. The 1-cocycles $R_{(a,n)}$, $R_{(n,b)}$, $R_{(n,0)}$ and $R_{(0,n)}$ represent in general non-trivial cohomology classes in $H^1(M_n; \mathbb{Z})$. $R_{(a,n)}$ and $R_{(n,b)}$ depend in general non-trivially on the parameter a respectively b. Moreover, $R_{(0,1)}$ for $n = 1$*

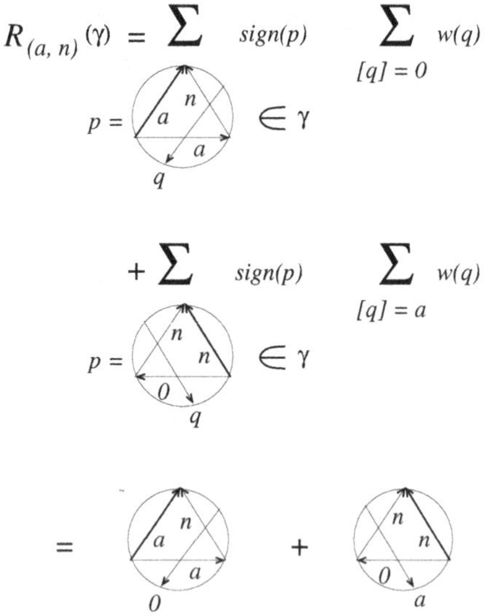

$$R_{(a,\,n)}(\gamma) = \sum_{p=} \text{sign}(p) \sum_{[q]\,=\,0} w(q)$$

$$+ \sum_{p=} \text{sign}(p) \sum_{[q]\,=\,a} w(q)$$

$$= \quad + $$

Figure 2.7: The 1-cocycle $R_{(a,n)}$

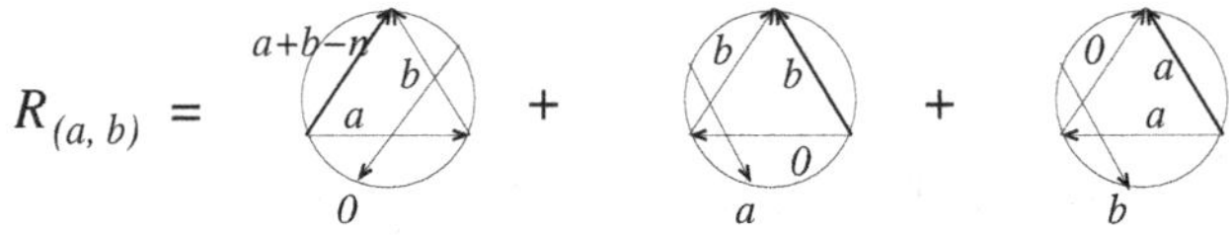

$$R_{(a,\,b)} = \quad + \quad + $$

Figure 2.8: The 1-cocycle $R_{(a,b)}$

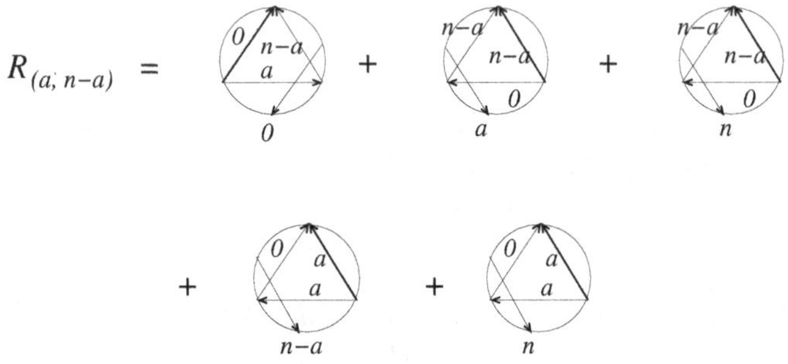

$$R_{(a,\,n-a)} = \quad + \quad + $$

$$+ \quad + $$

Figure 2.9: The 1-cocycle $R_{(a,n-a)}$

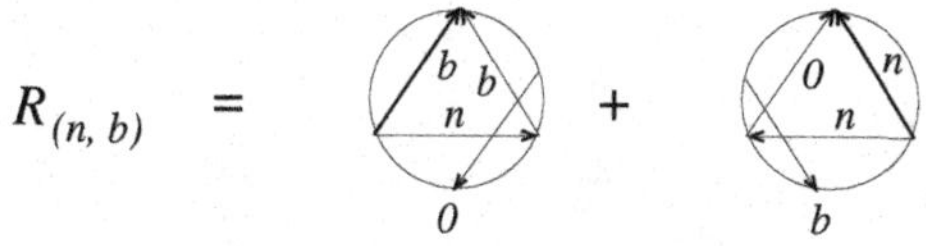

$$R_{(n,\,b)} \quad = \quad \text{} \quad + \quad \text{}$$

Figure 2.10: The 1-cocycle $R_{(n,b)}$

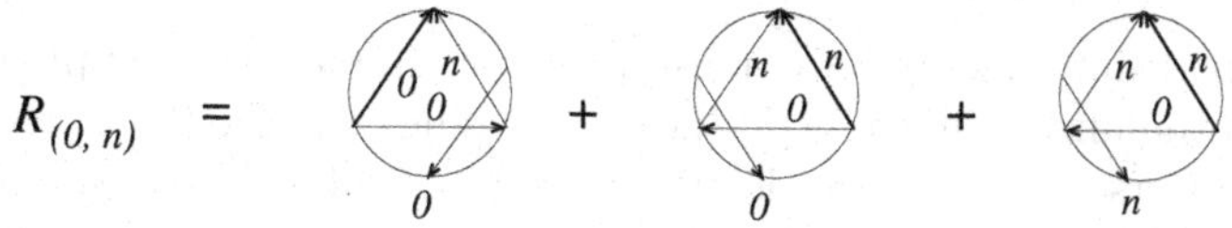

$$R_{(0,\,n)} \quad = \quad + \quad + \quad$$

Figure 2.11: The 1-cocycle $R_{(0,n)}$

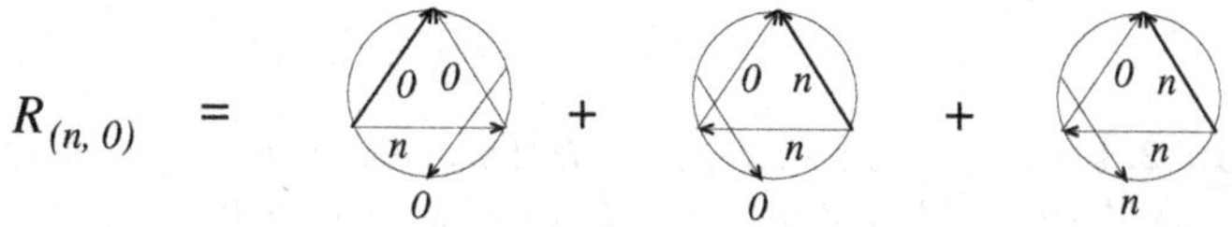

$$R_{(n,\,0)} \quad = \quad + \quad + \quad$$

Figure 2.12: The 1-cocycle $R_{(n,o)}$

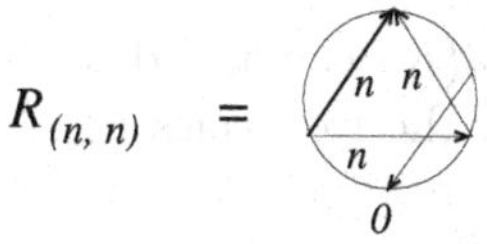

$$R_{(n,\,n)} \quad = \quad$$

Figure 2.13: The 1-cocycle $R_{(n,n)}$

Figure 2.14: The inverse of a knot K in the solid torus V

coincides with Mortier's 1-cocycle for long knots (which coincides probably with the Teiblum-Turchin 1-cocycle).

All our definitions are very non-symmetric. In fact, if we apply our definitions to the mirror image of K (i.e. all crossings switched, denoted by $K!$) or to the inverse knot (for the definition see Fig. 2.14, denoted by $-K$), then in general our 1-cocycles become new "dual" 1-cocycles.

We set $R_{(a,b)} = 0$ outside the closed 2-simplex.

Question 2.1 *Does there exist a $\mathbb{R}$-valued differential closed 1-form on M_n which depends on parameters $(a,b) \in \mathbb{R}^2$ and which coincides with our 1-cocycles for all integer parameters (a,b)?*

Definition 2.4 *$LR_{(a,n)}$ and $LR_{(n,b)}$ in $H^1(M_n; \mathbb{Z}[x])$ are defined as the Lagrange interpolation polynomials for $R_{(a,n)}$ respectively $R_{(n,b)}$ (i.e. the polynomial of the smallest degree which coincides with $R_{(a,n)}$ for $0 < a < n$).*

It is easy to see that $R_{(a,n)}(rot(K)) = -R_{(n,a)}(rot(-K))$ for each knot K. Indeed, if $K \in M_n$ then $-K \in M_n$ too. All signs and homological markings of crossings are preserved, but for Reidemeister III moves the crossings ml and hm are interchanged. Moreover the signs of the Reidemeister III moves change, because the loop $rot(K)$ is mapped to the loop $-rot(-K)$.

We do not know whether the 1-cocycles $R_{(a,b)}$ and $R_{(a,n-a)}$ are non-trivial too, because calculating interesting examples by hand is too difficult. Our few examples suggest the following question.

Question 2.2 *Is it true that $[R_{(a,n)}] = -[R_{(n,a)}]$ for each $0 < a < n$ and that $[R_{(n,n)}] = 0$ for each knot $K \in M_n$?*

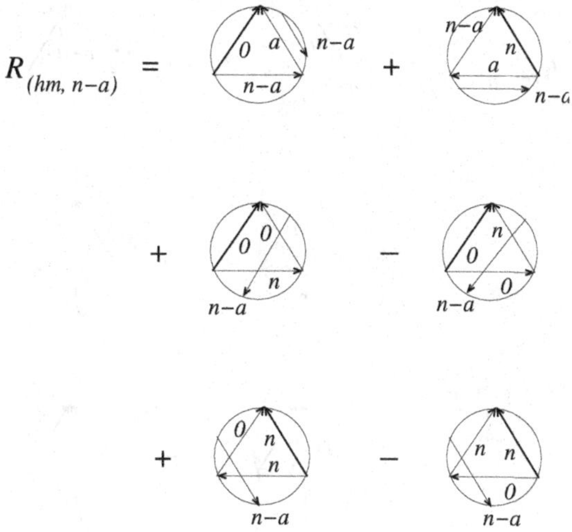

$$R_{(hm,\,n-a)} =$$

Figure 2.15: The 1-cocycle $R_{(hm,n-a)}$

Definition 2.5 *Let $n > 1$ and $0 < a < n$.*
The 1-cochain $R_{(hm,n-a)}$ in M_n is defined in Fig. 2.15.
The 1-cochain $R_{(ml,a)}$ in M_n is defined in Fig. 2.16.

Theorem 2.2 *The 1-cochains $R_{(hm,n-a)}$ and $R_{(ml,a)}$ are 1-cocycles in M_n for each knot K and each $0 < a < n$. They represent in general non-trivial cohomology classes in $H^1(M_n; \mathbb{Z})$.*

Notice that the diagrammatic formulas of the 1-cocycles from Theorem 2.2 contain also isolated arrows in contrast to the 1-cocycles from Theorem 2.1. Examples show that e.g. the 1-cocycles $R_{(ml,a)}$ and $R_{(a,n)}$ are in fact independent of each other.

Definition 2.6 *Let $n > 1$ and $0 < a < n$.*
The 1-cochain $R_{(a,a,n)}$ in M_n is defined in Fig. 2.17.

Theorem 2.3 *The 1-cochain $R_{(a,a,n)}$ is a 1-cocycle in M_n for each knot K and each $0 < a < n$. It represents in general a non-trivial cohomology class in $H^1(M_n; \mathbb{Z})$.*

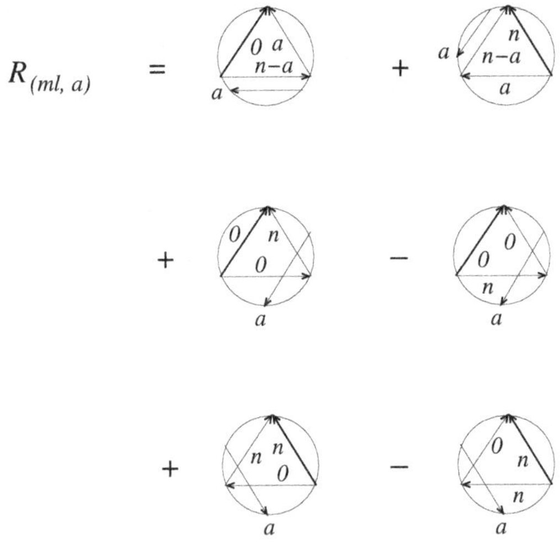

$$R_{(ml,\,a)} \quad = \quad \text{}$$

Figure 2.16: The 1-cocycle $R_{(ml,a)}$

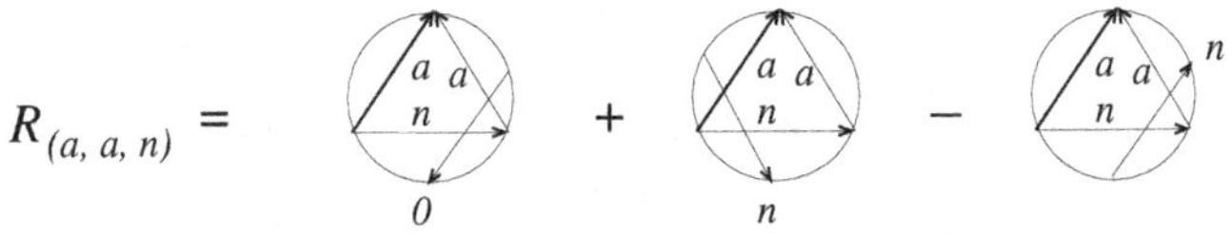

$$R_{(a,\,a,\,n)} = \quad \text{}$$

Figure 2.17: The 1-cocycle $R_{(a,a,n)}$

Notice that $R_{(a,a,n)}$ is defined by using only *one* global type of R III moves (or triple crossings) in contrast to the other 1-cocycles. We will see later that $R_{(a,a,n)}$ can be generalized straight forwardly to a 1-cocycle in M, i.e. the space of all knot diagrams without any restriction on negative loops, and not only in M_n, in contrast to all other of our 1-cocycles.

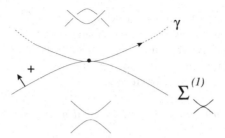

Figure 2.18: A loop γ is tangential from the positive side to a self-tangency

2.2 A quantization of $R_{(a,a,n)}$ in the free loop space

If we want to quantize a 1-cocycle, i.e. associate a monomial instead of an integer to a R III move, then we need evidently that the contribution of P_i cancels out with that of $\bar{P}_i$ for each $i \in \{1, 2, 3, 4\}$ in the positive global tetrahedron equations (compare the Introduction). It is hence a necessary condition that the diagrammatic 1-cocycle uses only one global type of triple crossings, as e.g. $R_{(a,n)}$.

Let $\mathbb{L}M_n$ be the free loop space of M_n and let $\bar{\mathbb{L}}M_n^{(1)}$ be the closure of all those strata of codimension 1 in $\mathbb{L}M_n$, which correspond to generic loops in M_n which are ordinary tangential to $\Sigma_{tan}^{(1)} \subset M_n$ from the positive side in exactly one point, compare Fig. 2.18.

Each 1-cocycle for M_n is a constant function on each component of $\mathbb{L}M_n$. We will define a polynomial valued 1-cocycle, but which is a constant function only on the components of $\mathbb{L}M_n \backslash \bar{\mathbb{L}}M_n^{(1)}$. In other words, it can change when a generic loop is deformed through a loop which touches $\Sigma_{tan}^{(1)}$ from the positive side.

Let γ be a generic oriented loop in M_n. We define the *trace graph* $TG(\gamma)$ of γ in Chapter 4, see also [20]. $\gamma : t \in S^1 \to M_n$ and we follow the crossings of $K \subset V$ in the isotopy of K represented by γ. $TG(\gamma)$ is a 6- and 1-valent graph in a thickened torus. The natural resolution of $TG(\gamma)$ is a disjoint union of circles and arcs. The arcs correspond always to crossings of marking 0 or n, because only these markings arise in R I moves (the end points of the arcs). For a fixed $0 < a < n$ we give names to to the circles which correspond to crossings of marking a: $a_1, a_2, \ldots, a_l$.

Each circle a_i has a natural projection $pr : a_i \to S^1 = \{t\}$ and we orient a_i in such a way that locally the degree of the projection pr is $+1$ if and only if the crossing is positive (pr has folds corresponding to the self-tangencies). If two crossings of marking a are born or die together in a R II move then they belong to the same component a_i.

Definition 2.7 $w(a_i) \in \mathbb{Z}$ *is defined as the degree of* $pr : a_i \to S^1 = \{t\}$.

Definition 2.8 (a, a_i, n) *is a R III move of global type* (a, a, n), *but such that the crossing hm belongs to the component* a_i.
$R_{(a,a_i,n)}$ *is the 1-cocycle from Definition 2.6, but where we sum up only over the moves* (a, a_i, n).

Notice that the crossing d has also the marking a, but it does not necessarily belong to the component a_i.

Definition 2.9 *Let* $n > 1$ *and* $0 < a < n$ *be fixed.*
The quantum 1-cocycle $qR_{(a,a,n)} \in \mathbb{Z}[x, x^{-1}]$ *is defined by*

$$qR_{(a,a,n)} = \sum_{i=1}^{l} w(a_i) x^{w(a_i) R_{(a,a_i,n)}}$$

where the sum is only over those components of a_i *which actually contain a crossing hm for some move* (a, a_i, n).

It could be that $R_{(a,a_i,n)} = 0$, but $w(a_i) \neq 0$. If a_i is never a crossing hm then we associate 0 to this component a_i. But if it arises as some crossing hm then we associate to it the constant $w(a_i) x^0$.

Proposition 2.1 $qR_{(a,a,n)}$ *is a 0-cocycle on* $\mathbb{L}M_n \setminus \bar{\mathbb{L}}M_n^{(1)}$ *and it represents in general a non-trivial cohomology class.*

The components of $TG(\gamma)$ change only if γ becomes tangential to $\Sigma_{tan}^{(1)}$. Tangencies on the negative side correspond to Morse modifications of $TG(\gamma)$ of index 0 and 2, but tangencies on the positive side correspond to Morse modifications of index 1 (compare Chapter 4) and we can no longer control $qR_{(a,a,n)}$.
It follows immediately from the definitions that if $|w(a_i)| \leq 1$ for each component a_i then $d(qR_{(a,a,n)})/dx(1) = R_{(a,a,n)}$.

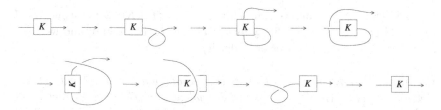

Figure 2.19: Nice realization of Gramain's loop

2.3 Examples

Let K be a long knot. Then Gramain's loop $rot(K)$ has a nice representative: with a Reidemeister I move we add a small positive curl with positive Whitney index at the right end, then we slide the knot along the curl, and at the end we eliminate the small curl again with a Reidemeister I move (compare Fig. 2.19 where we use a small positive curl with negative Whitney index).

Gramain's loop is still defined in the same way for M_n and for M, but the curl has to be replaced by its parallel n-cable. At the end we slide the curl in the solid torus with an isotopy of diagrams again in its initial position.

The Fox-Hatcher loop for long knots has a nice combinatorial realization too: we go on K from ∞ to the first crossing. If we arrive at an under-cross then we move the branch of the over-cross over the rest of the knot up to the end of K. If we arrive at an over-cross then we move the branch of the under-cross under the rest of the knot up to its end. We continue the process up to the moment when we obtain a diagram which is isotopic to our initial diagram of K, compare [29]. We can of course consider the analog loop for cables of long framed knots in M_n without moving the string link T which was used to close the parallel n-cable to a knot in V. It follows that the complexity of our 1-cocycles is of degree 3 with respect to the number of crossings for each fixed n. Indeed, by moving an arc over or under the rest of the diagram each of the c crossings contributes at most with a weight of degree 1 and we have to move at most $2c$ arcs in the loop.

Let (K', w) be a long framed knot. We introduce two new loops. First we replace K' it by its parallel n-cable nK' and close it with a string link T to a knot K in the solid torus V. We consider the loop $rot(nK')$, which is the rotation of nK' around the core of the solid torus but without moving T at all (then of course, at the end we have to push a full-twist of the bands back through nK'). Analogue, we can define the loop $fh(nK')$ without moving T (we could imagine T near infinity for the n-cable of a long knot and we never move bunches of strands over or under it).

Lemma 2.1 *If K' is a non-trivial torus knot then the non-trivial classes $[rot(nK')]$ and $[fh(nK')]$ are linearly dependent in $H_1(M_n; \mathbb{Q})$ for each $n > 0$ and each closure to a knot in the solid torus with a string link T.*

Proof. Hatcher [29] has proven that for $n = 1$ the moduli subspace of long knots $\subset M_1$ (i.e. no branch can ever move over infinity, or equivalently, there is a fixed compression disc for V which cuts the knot always in exactly one point) deformation retracts onto a circle and more precisely, a certain non-trivial integer multiple of the loop $rot(K')$ is homotopic to some non-trivial integer multiple of the loop $fh(K')$. (The homotopy class of $rot(K')$ does not depend on the framing and changing the framing adds multiples of $rot(K')$ to $fh(K')$.) We approximate $rot(K')$ and $fh(K')$ by loops without Reidemeister I moves in M_1 by using Whitney tricks (see e.g. [16]). We replace now the framed long knot by a n-cable which we can imagine on a band which projects to the plan by an immersion. The same generic homotopy as for the knot applies still to the band besides that a Reidemeister I move of the knot is now replaced by a Reidemeister I move of the band together with a negative full-twist of the band if the crossing from the cusp is positive and with a positive full-twist of the band if the crossing from the cusp is negative. There can of course occur Reidemeister I moves in the homotopy because a loop can become tangential to $\Sigma_{cusp}^{(1)}$ or can pass through $\Sigma_{trans-cusp}^{(2)}$ or $\Sigma_{cusp-deg}^{(2)}$ (compare Proposition 1.1). But the homotopy starts and ends with two loops of bands (with a common band as starting point) which all projects to the plane as immersions and which have all the same writhe = framing. It follows that the homotopy adds the same number of positive and negative full-twists to the bands. But full-twists can be moved along the bands (in a unique way up to homotopy because a full-twist is not allowed to move over T) and moving a full-twist commutes with any regular isotopy of a band. Therefore we can cancel out the positive and the negative full-twists two by two and we obtain a homotopy from a multiple of $rot(nK')$ to a multiple of $fh(nK')$ and the conclusion of the lemma follows. $\square$

Let R be any of our 1-cocycles $R_{(a,n)}$, $R_{(ml,a)}$, $R_{(a,a,n)}$. It follows from the lemma that if K' is a non-trivial torus knot then the rational function

$LR(fh(nK'))/LR(rot(nK')$ is in fact a rational number for each n and each string link T, such that $K \subset V$ is a knot. It would be very interesting to calculate examples of $LR(fh(nK'))/LR(rot(nK'))$ for K' the figure eight knot. Is it still a constant?

We give some very simple examples, just to show that our 1-cocycles are not always trivial.

Definition 2.10 *For a framed long knot K' and a n-component string link T, which induces a cyclic permutation of its end points, we denote by $push(T, K')$ the loop in M_n which consists of pushing T in the solid torus V once through the parallel n-cable of K'.*

The values of our 1-cocycles on these loops are isotopy invariants of framed classical knots for each fixed string link T.

Example 2.1 *Let K' be the standard diagram of the positive trefoil, $n = 3$ and let $T = \sigma_1\sigma_2$. We consider the loop $push(T, K')$.*
The two crossings in T have the homological markings 1 and 2. Consequently, there is never a Reidemeister III move with two crossings of marking 0 or 3 in the triangle. It follows that

$$R_{(0,3)}(push(T, K')) = R_{(3,0)}(push(T, K')) = 0.$$

It follows also that in $R_{(1,3)}(push(T, K'))$ only the first configuration in the formula could contribute non-trivially. One easily sees that there is only a single R III move in the loop which is of global type $([d] = 1, [hm] = 3, [ml] = 1)$. We show this move in Fig. 2.20. In the figure we indicate all crossings of homological marking 0. The R III move has sign $= +1$ and there is exactly one crossing which contributes to the weight of $R_{(1,3)}$. For the convenience of the reader we encircle the contributing crossings in the figures. The solid torus is in all figures the complement of the z-axes in 3-space, which is indicated by a point in the plan.
It follows that

$$R_{(1,3)}(push(T, K')) = 1.$$

An easy calculation gives

$$R_{(3,1)}(push(T, K')) = R_{(1,1,3)}(push(T, K')) = -1.$$

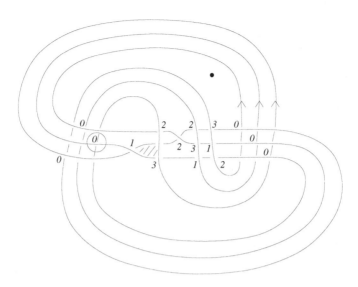

Figure 2.20: The only contribution to $R_{(1,3)}(push(T, K'))$

For the same reason in the calculation of $R_{(hm,n-2=1)}(push(T, K'))$ only the first two configurations could contribute non-trivially. We show their unique contributions in Fig. 2.21, where we indicate all homological markings and we encircle the unique crossings which contribute to the weights. The signs of both R III moves are -1. It follows that

$$R_{(hm,1)}(push(T, K')) = -2.$$

The following is an extremely simple example in order to show that the 1-cocycles depend in general on the integer parameter a.

Example 2.2 *Let K' be again the standard positive trefoil, $n = 4$ and $T = \sigma_1 \sigma_2^{-1} \sigma_3$. The homological markings in T are correspondingly 1, 2, and 3. A calculation completely analogue to that in Example 2.1, see Fig. 2.22, shows that*

$$R_{(1,4)}(push(T, K')) = R_{(3,4)}(push(T, K')) = 1.$$

We show the unique R III move which could contribute to $R_{(2,4)}(push(T, K'))$ in Fig. 2.23. Its sign is -1 and it follows that

$$R_{(2,4)}(push(T, K')) = -1.$$

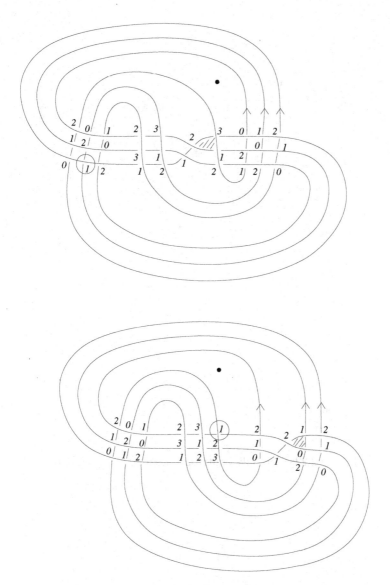

Figure 2.21: The only contributions to $R_{(hm,n-2=1)}(push(T, K'))$

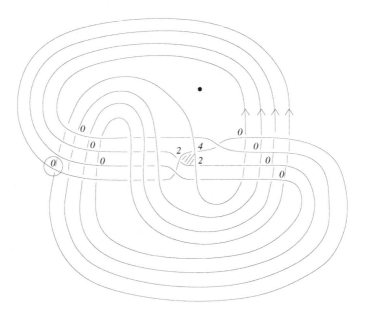

Figure 2.22: The only contribution to $R_{(2,4)}(push(T, K'))$

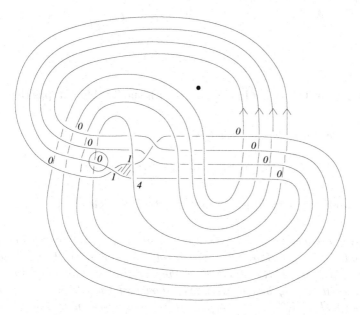

Figure 2.23: The only contribution to $R_{(1,4)}(push(T, K'))$

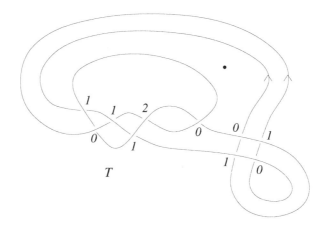

Figure 2.24: The string link T closed with the 2-cable of a positive curl

A standard calculation gives now

$$LR_{(a,4)}(push(T, K')) = 2x^2 - 8x + 7.$$

T is no longer a braid in the following two examples.

Example 2.3 *Let K' be the trivial knot with framing $+1$, $n = 2$ and T the 2-string link shown in Fig. 2.24. We consider $push(T, K'))$ which is evidently the same loop as $rot(K)$, where K is the trivial closure of $T\sigma_1^{-2}$ in V. On easily sees that only one R III move, shown in Fig. 2.25, can contribute to $R_{(0,2)}(push(T, K'))$ and only one R III move, shown in Fig. 2.26, can contribute to $R_{(1,2)}(push(T, K'))$. It follows that*

$$R_{(0,2)}(push(T, K')) = -1 \text{ and } R_{(1,2)}(push(T, K')) = -1 \text{ too.}$$

A calculation gives

$$R_{(1,1,2)}(push(T, K')) = 1 \text{ and } R_{(hm,1)}(push(T, K')) = -2.$$

Example 2.4 *Let K' be the standard diagram of the positive trefoil, $n = 2$ and T the same 2-string link shown in Fig. 2.27.*

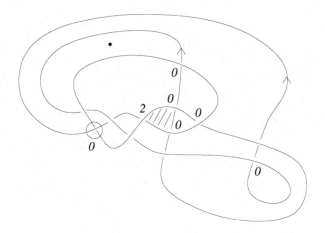

Figure 2.25: The only contribution to $R_{(0,2)}(push(T, K'))$

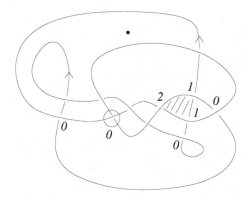

Figure 2.26: The only contribution to $R_{(1,2)}(push(T, K'))$

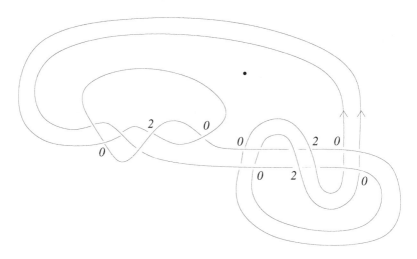

Figure 2.27: The string link T closed with the 2-cable of the positive trefoil

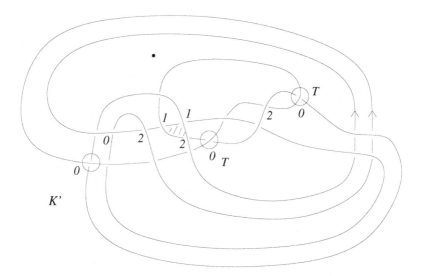

Figure 2.28: Contributions to the weight of a R III move, which come from different parts of the knot

A bit longer calculations give

$$R_{(0,2)}(push(T, K')) = -3$$

$$R_{(1,2)}(push(T, K')) = -2$$

$$R_{(1,1,2)}(push(T, K')) = 2$$

$$R_{(hm,1)}(push(T, K')) = -8.$$

An important feature is the fact, that our 1-cocycles mix the contributions to the weights from crossings of T with those from crossings of K' in the same R III move! We show an example for $R_{(1,2)}(push(T, K'))$ in Fig. 2.28. Hence, they seem to make a non-trivial connection between T and K'.

It follows immediately from the above examples that $R_{(0,n)}$, $R_{(a,n)}$ and $R_{(hm,n-a)}$ represent linearly independent 1-cohomology classes. But it could well be that $[R_{(n,a)}] = [R_{(a,a,n)}]$. However, we will see that $R_{(a,a,n)}$ is still well defined in M if we reduce the homological markings to $\mathbb{Z}/n\mathbb{Z}$ but $R_{(n,a)}$ is not.

Example 2.5 *Let K' be the standard diagram of the figure eight knot shown in Fig. 2.29. We take $n = 2$ and $T = \sigma_1$. Evidently, a R III move is of type $([d] = 1, [hm] = 1, [ml] = 2)$ only when T moves over a crossing of K' such that in the 2-cable there is a crossing of marking 2 generated by the crossing of K'. In this case the crossing hm is always just the crossing in T. It follows (see Fig. 2.29) that*

$$qR_{(1,1,2)}(push(T, K')) = x.$$

Let $K'!$ be the mirror image of K' (which is framed isotopic to K'). A bit longer calculation shows that

$$qR_{(1,1,2)}(push(T, K'!)) = x \text{ too.}$$

The two loops are of course homotopic, but we do not know whether they are actually in the same component of the free loop space $\mathbb{L}M_2 \setminus \bar{\mathbb{L}}M_2^{(1)}$.

2.4 Proofs

Our main results are based on rather complicated combinatorics together with singularity theory.

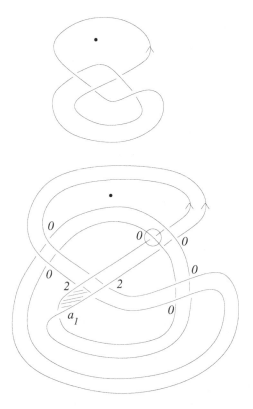

Figure 2.29: The only contribution to $qR_{(1,1,2)}(push(T, K'))$

2.4.1   Generalities and reductions by using singularity theory

We consider the moduli space M of all oriented knots in V and with a fixed projection $pr : V \to S^1 \times \mathbb{R}$. A generic loop s in M intersects $\Sigma_{tri}^{(1)}$, $\Sigma_{tan}^{(1)}$ and $\Sigma_{cusp}^{(1)}$ transversely in a finite number of points and it does not intersect at all strata of higher codimension. To each intersection point with $\Sigma_{tri}^{(1)}$ we associate a contribution in $\mathbb{Z}$ for a fixed 1-cochain R. We sum up with signs (coming from the co-orientation) the contributions over all intersection points in s and we obtain $R(s)$.

We use now the strata from $\Sigma^{(2)}$ to show the invariance of $R(s)$ under all generic homotopies of s in M. A homotopy is *generic* if it intersects $\Sigma^{(1)}$ transversely besides for a finite number of points in s where s has an ordinary tangency with $\Sigma^{(1)}$, it intersects $\Sigma^{(2)}$ transversely in a finite number of points and it doesn't intersect at all $\Sigma^{(i)}$ for $i > 2$. We see immediately that $R(s)$ is invariant by passing through a tangency of s with $\Sigma^{(1)}$. Indeed, the two intersection points have identical contributions but they enter with different signs and cancel out. In order to show the invariance under generic homotopies we have to study now normal 2-discs for the strata in $\Sigma^{(2)} \subset M$. For each type of stratum in $\Sigma^{(2)}$ we have to show that $R(m) = 0$ for the boundary m of the corresponding normal 2-disc in M. We call m a *meridian*. $\Sigma^{(2)}_{quad}$ is by fare the hardest case which leads to the tetrahedron equation. We look at the tetrahedron equation from the point of view of singularities of the projection of lines as explained in the Introduction.

Going along the meridian m of the positive quadruple crossing we see ordinary diagrams of positive 4-braids and exactly eight diagrams with an ordinary positive triple crossing. We show this in Fig. 2.30. (For simplicity we have drawn the positive triple crossings as triple points, but the branches do never intersect.) However, we have to study 48 different local types of quadruple crossings for each global type of a quadruple crossing.

It is convenient to make a first coarse distinction of global types of triple crossings: *type r* if the arrow which corresponds in the triangle to the crossing ml goes to the right, and *type l* if it goes to the left. The different local types of triple crossings are shown and numbered in Fig. 2.31. The sign in the figure indicates the side of the discriminant $\Sigma^{(1)}$ if the triple crossing is of global type r. If it is of global type l then all signs have to be changed to the opposite ones.

Triple crossings come together in points of $\Sigma^{(2)}_{trans-self}$, but one easily sees that the global type of the triple crossings (i.e. its Gauss diagram with the homological markings but without the writhe) is always preserved. We make now a graph Γ for each global type of a triple crossing in the following way: the vertices correspond to the different local types of triple crossings. We connect two vertices by an edge if and only if the corresponding strata of triple crossings are adjacent to a stratum of $\Sigma^{(2)}_{trans-self}$. One easily sees that the resulting graph is the 1-skeleton of the 3-dimensional cube I^3 (compare Section 2.4.3). In particular, it is connected. (Studying the normal discs to $\Sigma^{(2)}_{trans-self}$ in M one shows that if a 0-cochain is invariant under passing

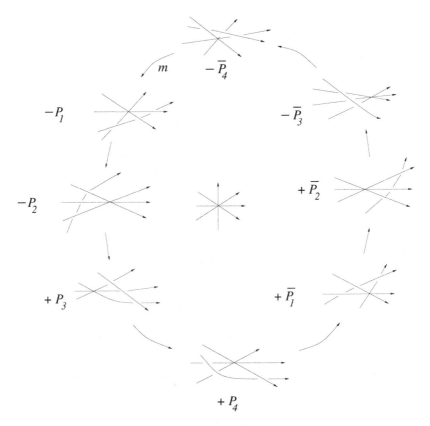

Figure 2.30: Unfolding of a positive quadruple crossing

all $\Sigma_{tan}^{(1)}$ and just one local type of a stratum $\Sigma_{tri}^{(1)}$ then it is invariant under passing all remaining local types of triple crossings because Γ is connected.) The edges of the graph $\Gamma = skl_1(I^3)$ correspond to the types of strata in $\Sigma_{trans-self}^{(2)}$. The solution of the positive tetrahedron equation tells us what is the contribution to R of a positive triple crossing (i.e. all three involved crossings are positive). The meridians of the strata from $\Sigma_{trans-self}^{(2)}$ give equations which allow us to determine the contributions of all other types of triple crossings. However, a global phenomenon occurs: each loop in Γ could give an additional equation. Evidently, it suffices to consider the loops which are the boundaries of the 2-faces from $skl_2(I^3)$. We call all the

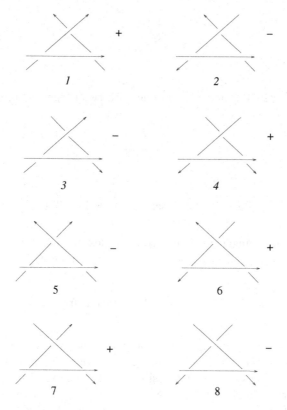

Figure 2.31: Local types of a triple crossing

equations which come from the meridians of $\Sigma^{(2)}_{trans-self}$ and from the loops in $\Gamma = skl_1(I^3)$ the *cube equations* (Section 2.4.3). (Notice that a loop in Γ is more general than a loop in M. For a loop in Γ we come back to the same local type of a triple crossing but not necessarily to the same whole diagram of the knot.)

We need only the following strata from $\Sigma^{(3)}$ in order to simplify the proof of the invariance of R in generic homotopies which pass through strata from $\Sigma^{(2)}$:

- (1) one degenerate quadruple crossing where exactly two branches have an ordinary self-tangency in the quadruple point, denoted by $\Sigma^{(3)}_{trans-trans-self}$ (see Fig. 2.32).

Figure 2.32: A quadruple crossing with two tangential branches

Figure 2.33: A self-tangency in a flex

- (2) one self-tangency in an ordinary flex with a transverse branch passing through the tangent point, denoted by $\Sigma^{(3)}_{trans-self-flex}$ (see Fig. 2.33).

- (3) the transverse intersection of a stratum from $\Sigma^{(1)}$ with a stratum of $\Sigma^{(2)}_{trans-self}$.

Again, for each fixed global type of a quadruple crossing we form a graph with the local types of quadruple crossings as vertices and the adjacent strata of $\Sigma^{(3)}_{trans-trans-self}$ as edges. One easily sees that the resulting graph Γ has exactly 48 vertices and that it is again connected. Luckily, we don't need to study the unfolding of $\Sigma^{(3)}_{trans-trans-self}$ in much detail. It is clear that each meridional 2-sphere for $\Sigma^{(3)}_{trans-trans-self}$ intersects $\Sigma^{(2)}$ transversely in a finite number of points, namely exactly in two strata from $\Sigma^{(2)}_{quad}$ and in lots of strata from $\Sigma^{(2)}_{trans-self}$ and from $\Sigma^{(1)} \cap \Sigma^{(1)}$ and there are no other intersections with strata of codimension 2. If we know now that $R(m) = 0$ for the meridian m of one of the quadruple crossings, that $R(m) = 0$ for all meridians of $\Sigma^{(2)}_{trans-self}$ (i.e. R satisfies the cube equations) and for all meridians of $\Sigma^{(1)} \cap \Sigma^{(1)}$ (i.e. R is invariant under passing simultaneous Reidemeister moves at different places in the diagram) then $R(m) = 0$ for the other quadruple crossing too. It follows that for each of the fixed global types (see Fig. 2.34, where in each case we have to consider all possible markings of the six arrows) the 48 tetrahedron equations reduce to a single one, which is called the *positive global tetrahedron equation*. There is no further reduction possible because

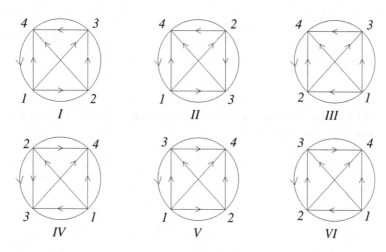

Figure 2.34: The global types of quadruple crossings

we are searching for non symmetric solutions and which depend non-trivially on the markings!

In the cube equations there are also two local types of edges, corresponding to the the two different local types of a Reidemeister II move with given orientations, compare Fig. 2.35. We reduce them to a single type of the edge by using the strata from $\Sigma^{(3)}_{trans-self-flex}$. The meridional 2-sphere for $\Sigma^{(3)}_{trans-self-flex}$ intersects $\Sigma^{(2)}$ transversely in exactly two strata from $\Sigma^{(2)}_{trans-self}$, which correspond to the two different types of the edge, and in lots of strata from $\Sigma^{(2)}_{self-flex}$ and from $\Sigma^{(1)} \cap \Sigma^{(1)}$. Hence, again the invariance under passing one of the two local types of strata from $\Sigma^{(2)}_{trans-self}$ together with the invariance under passing all strata from $\Sigma^{(2)}_{self-flex}$ and $\Sigma^{(1)} \cap \Sigma^{(1)}$ guaranties the invariance under passing the other local type of $\Sigma^{(2)}_{trans-self}$ too.

The unfolding (i.e. the intersection of a normal disc with the stratification of M) of e.g. the edge $1 - 5$ of Γ is shown in Fig. 2.36 (compare [20] and Chapter 4). The intersection of a meridional 2-sphere for $\Sigma^{(1)} \cap \Sigma^{(2)}_{trans-self}$ with $\Sigma^{(2)}$ consists evidently of the transverse intersections of the stratum $\Sigma^{(1)}$ with all the strata of codimension 1 in the unfolding of $\Sigma^{(2)}_{trans-self}$. It follows again that it is sufficient to prove the invariance under passing $\Sigma^{(1)} \cap \Sigma^{(1)}$

Figure 2.35: Two different local types of an edge 1-7 in the cube equations and which come together in $\Sigma^{(3)}_{trans-self-flex}$

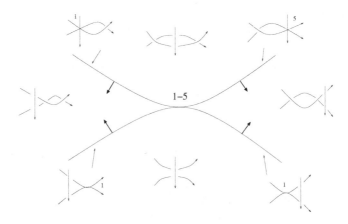

Figure 2.36: The unfolding of a self-tangency with a transverse branch corresponding to the edge $1 - 5$

only for positive triple crossings and only for one of the two local types of self-tangencies with a given orientation.

Because we consider only contributions from R III moves, it follows immediately that $R(m) = 0$ on meridians of the strata of the form (3) and (6) in Proposition 1.1 (compare Fig. 2.37 and Fig. 2.38). Because we consider only linear weights, i.e. only a single arrow outside of the triangle, $R(m) = 0$ for (4), in the case of $\Sigma^{(1)}_{tri} \cap \Sigma^{(1)}_{tri}$ (compare Fig. 2.39). In the case of $\Sigma^{(1)}_{tri} \cap \Sigma^{(1)}_{tan}$ the two new crossings from the self-tangency cancel out together in the weight. In the case $\Sigma^{(1)}_{tri} \cap \Sigma^{(1)}_{cusp}$ there appears a new arrow of marking 0 or n, but which does not intersect the triangle. In all our 1-cocycles R there is never a weight of this type, and hence they are 0 on this meridian too.

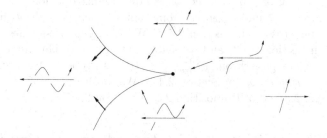

Figure 2.37: The unfolding for the self-tangency in a flex

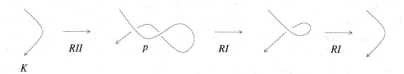

Figure 2.38: A meridian for a degenerated cusp

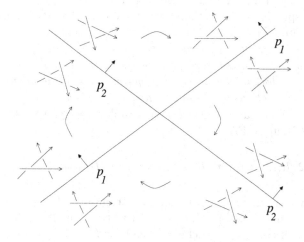

Figure 2.39: Meridian of two simultaneous Reidemeister III-moves

The remaining part of this section is now organized as follows: we show the invariance of R under generic homotopies which pass through a positive quadruple crossing in Section 2.4.2. We show that R satisfies the cube equations in Section 2.4.3, and we show that R is invariant under generic homotopies which pass through $\Sigma^{(2)}_{trans-cusp}$ ((5) in Proposition 1.1 and which we call *wandering cusps*) in Section 2.4.4. We finish the proof of the Theorems 2.1, 2.2, 2.3 and of Proposition 2.1 in Section 2.4.5.

2.4.2   The positive global tetrahedron equation in M_n

This section contains the heart of this chapter.

As was explained in Section 2.4.1 it suffices to consider global positive quadruple crossings. We naturally identify crossings in an isotopy outside Reidemeister moves of type I and II. The Gauss diagrams of the unfoldings of the quadruple crossings are given in Fig. 2.40 up to Fig. 2.51. For the convenience of the reader (and for further research) we marque also the different possibilities for the point at infinity in the case of long knots, i.e. M_1. The (positive) crossing between the local branch i and the local branch j is always denoted by ij.

Using these figures we show in Fig. 2.52 up to Fig. 2.63 the homological markings of the arrows. Here α, β and γ are the homology classes represented by the corresponding arcs in the circle (remember that the circle is always oriented counter-clockwise). The ordering of the global types here is more adapted to the check for our 1-cocycles.

In M_n we have the unknown parameters α, β and γ in $\{0, 1, ..., n\}$ and $\alpha + \beta + \gamma \leq n$, because there are no negative loops in the diagrams. In M there are no restrictions at all on the integers α, β and γ.

Let m be a meridian of a positive quadruple crossing. We use now extensively Fig. 2.52 up to Fig. 2.63.

Lemma 2.2 $R(m) = 0$ *for all 1-cochains from Definition 2.3.*

Proof. We have only to study the contributions of the three crossings from the quadruple crossing which are not in the triangle, because all other contributions of crossings cancel always out for P_i and $\bar{P}_i$. First we observe, that the configurations used in the definitions of all the 1-cochains of Definition 2.3 occur only for the global types IV and II.

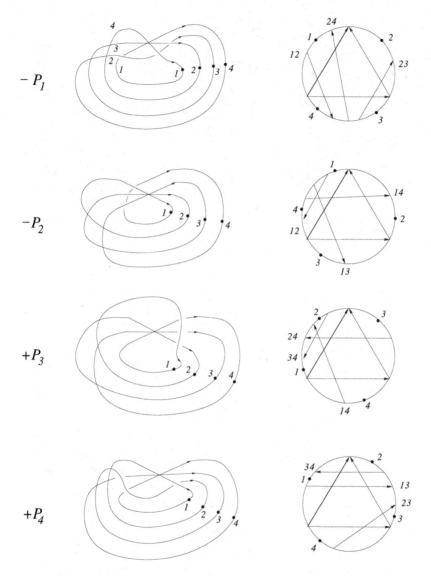

Figure 2.40: First half of the meridian for global type I

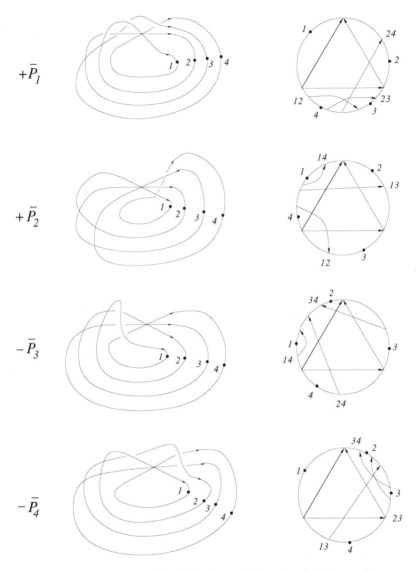

Figure 2.41: Second half of the meridian for global type I

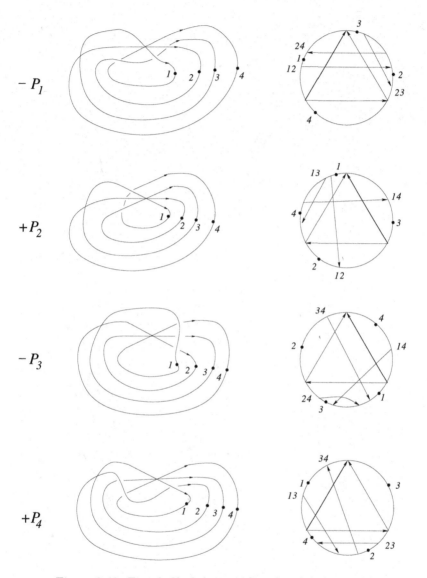

Figure 2.42: First half of the meridian for global type II

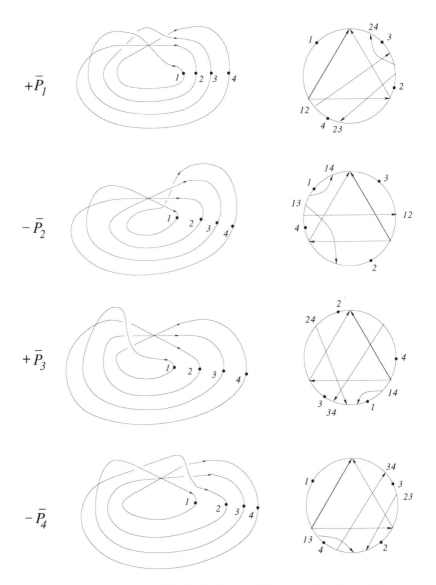

$+\bar{P}_1$

$-\bar{P}_2$

$+\bar{P}_3$

$-\bar{P}_4$

Figure 2.43: Second half of the meridian for global type II

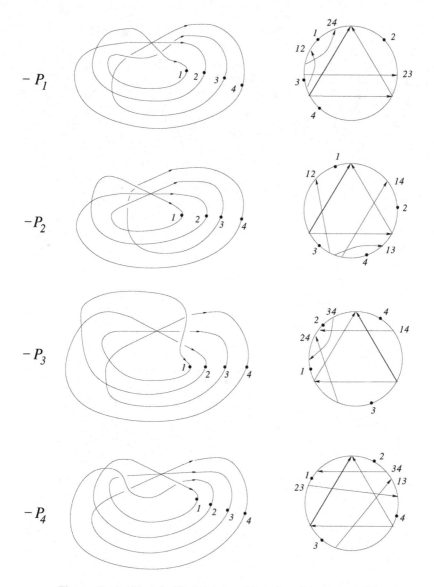

Figure 2.44: First half of the meridian for global type III

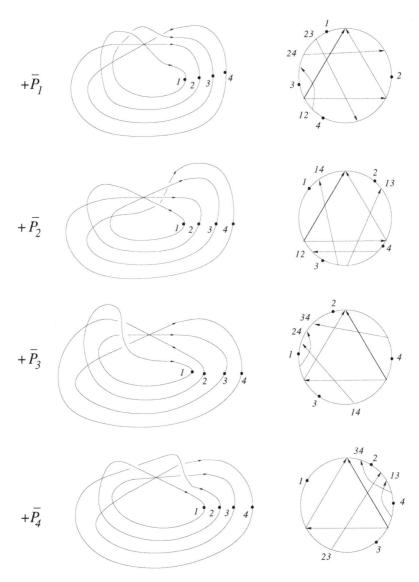

Figure 2.45: Second half of the meridian for global type III

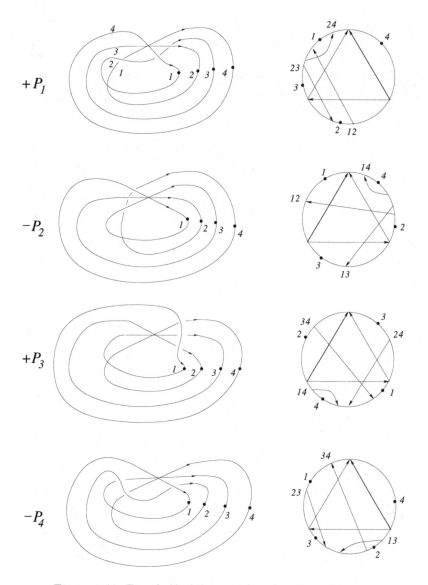

Figure 2.46: First half of the meridian for global type IV

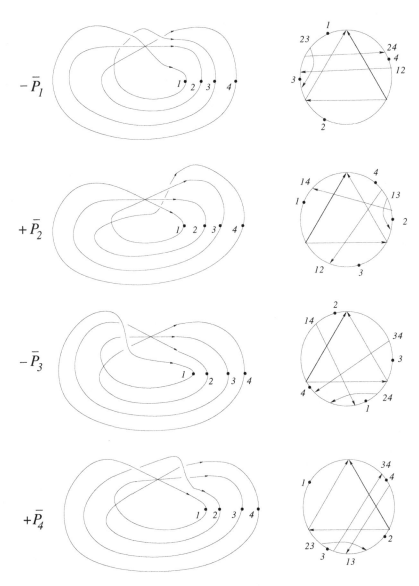

Figure 2.47: Second half of the meridian for global type IV

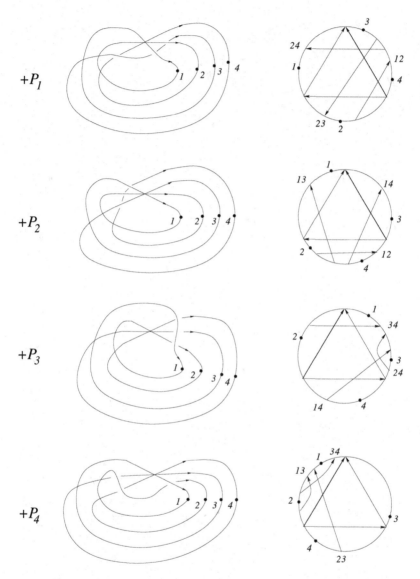

Figure 2.48: First half of the meridian for global type V

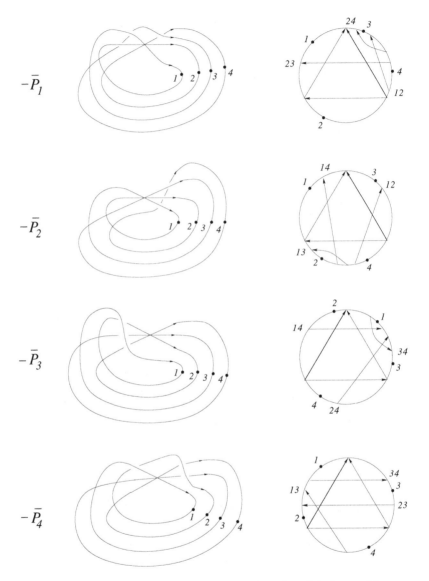

Figure 2.49: Second half of the meridian for global type V

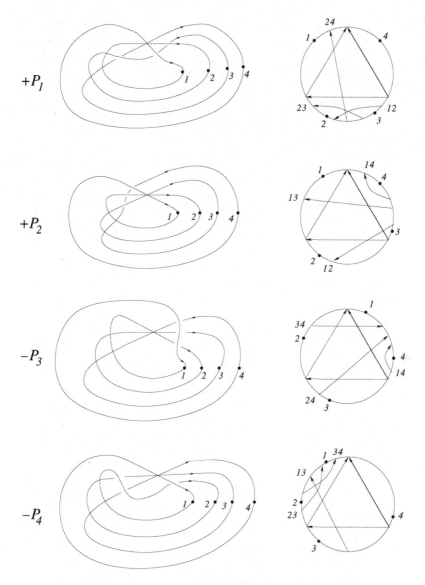

Figure 2.50: First half of the meridian for global type VI

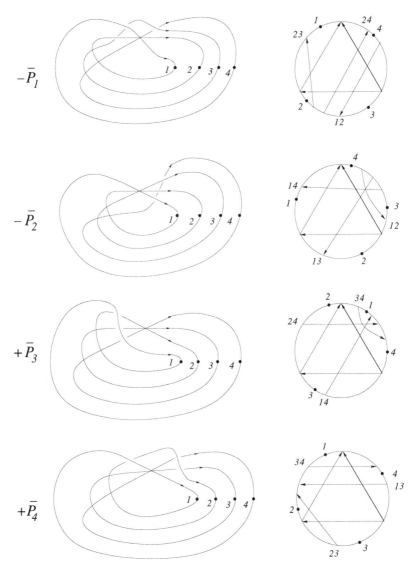

Figure 2.51: Second half of the meridian for global type VI

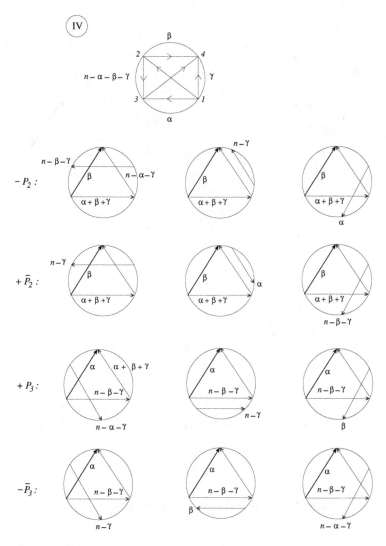

Figure 2.52: Moves of type r for the global type IV

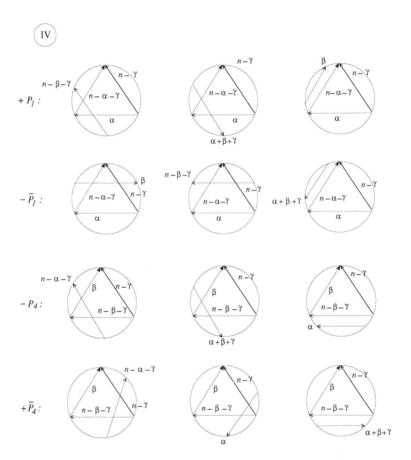

Figure 2.53: Moves of type l for the global type IV

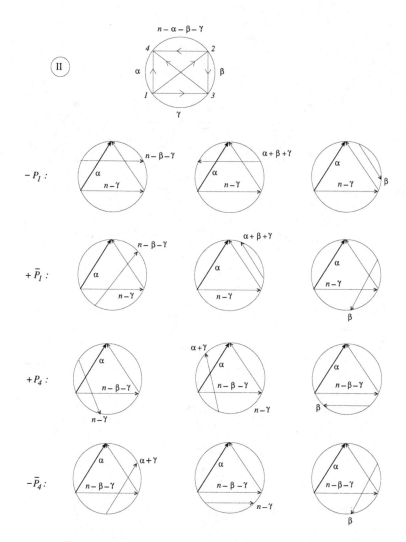

Figure 2.54: Moves of type r for the global type II

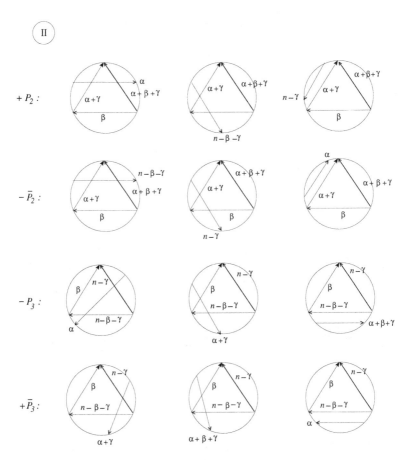

Figure 2.55: Moves of type l for the global type II

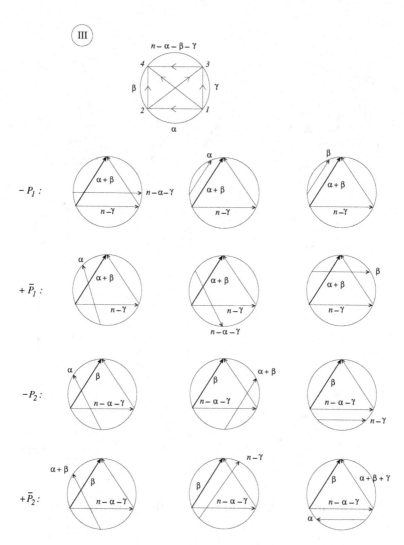

Figure 2.56: Moves of type r for the global type III

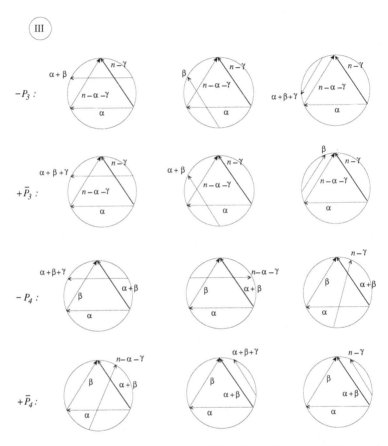

Figure 2.57: Moves of type l for the global type III

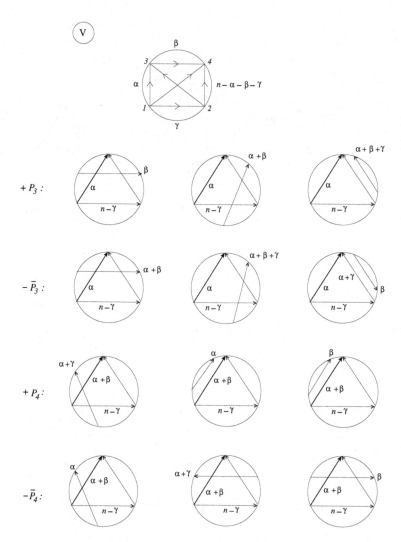

Figure 2.58: Moves of type r for the global type V

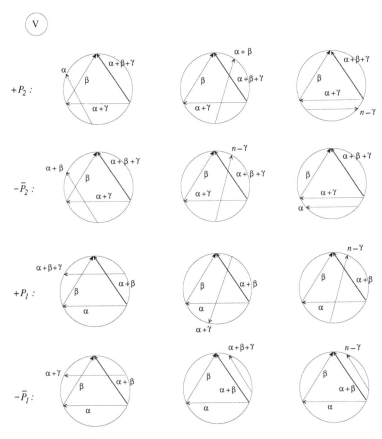

Figure 2.59: Moves of type l for the global type V

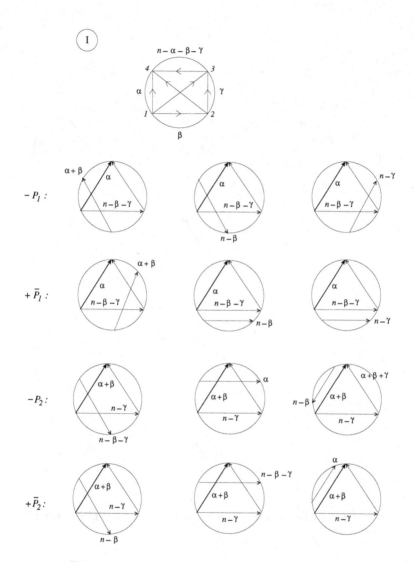

Figure 2.60: Moves of type r for the global type I

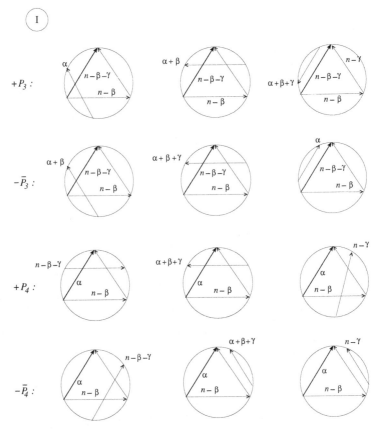

Figure 2.61: The remaining moves of type r for the global type I

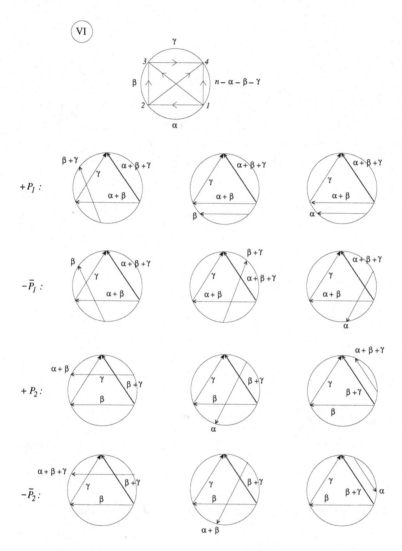

Figure 2.62: Moves of type l for the global type VI

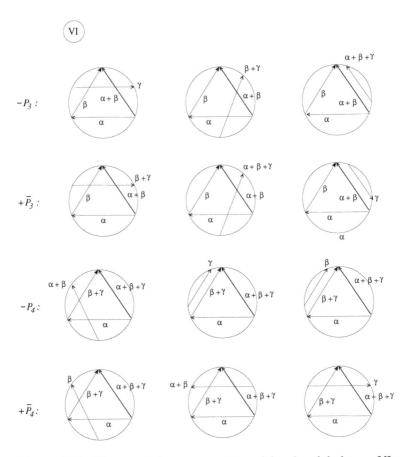

Figure 2.63: The remaining moves of type l for the global type VI

$R_{(a,b)}$:

Case IV:

$-P_2$ contributes if and only if $\alpha = 0$, $\beta = a + b - n$, and $\gamma = n - b$. Exactly in this case P_1 contributes too (with the second configuration in the formula).

If $\bar{P}_2$ contributes then $\beta + \gamma = n$ and $\alpha + \beta + \gamma = a$. But this can not happen in M_n because $\alpha \geq 0$.

P_3 contributes if and only if $\alpha = a + b - n$, $\beta = 0$ and $\gamma = n - a$. Exactly in this case $-P_4$ contributes with the third configuration too.

If $\bar{P}_3$ contributes then $\alpha = a + b - n$, $\gamma = 2n - a - b$ and $\beta = b - n$, which can not happen in M_n (remember that we consider the open 2-simplex here, i.e. $b < n$).

There are no configurations at all in $-\bar{P}_1$ and in $\bar{P}_4$ which could contribute.

Case II:

$\bar{P}_1$ contributes if and only if $\alpha = a + b - n$, $\beta = 0$ and $\gamma = n - a$. But exactly in this case $-\bar{P}_4$ contributes too.

If P_2 contributes with the second configuration then $\beta = 0$ and hence it cancels out with the contribution of $-\bar{P}_2$. If P_2 would contribute with the third configuration then $\beta = a$, $\alpha + \gamma = 0$ and hence in M_n we have $\alpha = 0$ and $\gamma = 0$. Hence $n - \beta - \gamma = b$ implies $a + b = n$. Impossible, because we are in the open 2-simplex. If $-\bar{P}_2$ would contribute with the third configuration then $n = b$, which is also impossible.

$-P_3$ contributes with the third configuration if and only if $\beta = 0$, $n - \gamma = a$ and $\alpha + \gamma = b$. But it cancels out with the contribution of $\bar{P}_3$. If $-P_3$ would contribute with the second configuration then $\beta = b$, $\gamma = n - b$ and $\alpha = a + b - n$. But then e.g. $\bar{P}_3$ contains an arrow with marking $\alpha + \beta + \gamma = a + b > n$, which is not possible in M_n. If $\bar{P}_3$ would contribute with the second configuration then $\beta = b$, $\gamma = n - b$ and $\alpha = a - n$, which is impossible.

This finishes the proof for $R_{(a,b)}$.

$R_{(a,n-a)}$:

Case IV:

$-P_2$ contributes if and only if $\alpha = \beta = 0$ and $\gamma = a$. But then it cancels out with the second configuration of the formula in P_1. If $\bar{P}_2$ would contribute then $\beta = 0$ and $\gamma = n$, but this would imply $a = 0$ in M_n. Not possible, because we are in the open 1-simplex.

P_3 contributes if and only if $\alpha = \beta = 0$ and $\gamma = n - a$. But then $-P_4$ contributes with the forth configuration. If $-\bar{P}_3$ would contribute then $\alpha = 0$ and $\gamma = n$, which is not possible in M_n for $a \neq 0$.

P_1 could contribute with $\gamma = a$, $\alpha = 0$ and $\beta = n - a$. Then it cancels out with $-P_4$. P_1 could contribute with $\gamma = n - a$, $\alpha = a$ and $\beta = 0$. It cancels out again with $-P_4$.

Case II:

$\bar{P}_1$ contributes if and only if $\alpha = 0$, $\gamma = n - a$ and $\beta = 0$. But then it cancels out with $-\bar{P}_4$.

If $\beta = 0$ in P_2 then it always cancels out with $-\bar{P}_2$. The same is true for $-P_3$ and $\bar{P}_3$. P_2 contributes if $\beta = a$ and $\alpha + \gamma = 0$. But then $\alpha = 0$ and $\gamma = 0$. It follows again that P_2 cancels out with $-\bar{P}_2$. $-P_3$ could contribute if $\beta = n - a$, $\gamma = a$ and $\alpha = 0$. In this case again it cancels out with $\bar{P}_3$.

This finishes the proof for $R_{(a,n-a)}$.

$R_{(n,b)}$:

Case IV:

$-P_2$ contributes if and only if $\beta = b$, $\alpha = 0$ and $\gamma = n - b$. But then it cancels out with $\bar{P}_2$.

P_3 contributes if and only if $\alpha = b$, $\beta = 0$ and $\gamma = 0$. Then it cancels out with $-P_4$. If $-\bar{P}_3$ would contribute, then $\alpha = b$, $\gamma = n - b$ and hence $\beta = b - n$, which is not possible in M_n.

If P_1 would contribute, then $\alpha = n$, $\gamma = 0$ and hence again $\beta = b - n$.

Case II:

$\bar{P}_1$ contributes if and only if $\alpha = b$ and $\gamma = \beta = 0$. But then it cancels out with $-\bar{P}_4$.

If P_2 would contribute, then $\beta = n$, $\alpha = b$ and $\gamma = -b$. Not possible in M_n. If $-\bar{P}_2$ would contribute, then $\beta = n$, $\gamma = n - b$ and $\alpha = b - n$. Not possible in M_n.

$-P_3$ contributes if and only if $\beta = \gamma = 0$ and $\alpha = b$. In this case it cancels out with $\bar{P}_3$.

This finishes the proof for $R_{(n,b)}$. The considerations for the 1-cochain $R_{(a,n)}$ are completely analogous and are left to the reader.

$R_{(0,n)}$:

Case IV:

$-P_2$ contributes if and only if $\alpha = \beta = \gamma = 0$. Then it cancels out with P_1. $\bar{P}_2$ never contributes.

P_3 contributes if and only if $\alpha = \beta = 0$ and $\gamma = n$. Then it cancels out with $-\bar{P}_3$.

P_1 could contribute also for $\alpha = \gamma = 0$ and $\beta = n$. Then it cancels out with $-P_4$.

Case II:

$\bar{P}_1$ contributes if and only if $\alpha = \beta = 0$ and $\gamma = n$. Then it cancels out with $-\bar{P}_4$.

P_2 can only contribute for $\beta = 0$ and cancels always out with $-\bar{P}_2$.

$-P_3$ can only contribute for $\beta = n$ and hence $\alpha = \gamma = 0$. But then it cancels out with $\bar{P}_3$.

This finishes the proof for $R_{(0,n)}$. The considerations for the 1-cochain $R_{(n,0)}$ are completely analogous and are left to the reader.

$R_{(n,n)}$:

Case IV:

$-P_2$ can only contribute for $\alpha = \gamma = 0$ and $\beta = n$. Then it cancels out with $\bar{P}_2$.

P_3 can only contribute for $\beta = \gamma = 0$ and $\alpha = n$. Then it cancels out with $-\bar{P}_3$.

Case II:

$\bar{P}_1$ can only contribute for $\beta = \gamma = 0$ and $\alpha = n$. But then it cancels out with $-\bar{P}_4$.

This finishes the proof for $R_{(n,n)}$. $\square$

Let $0 < b < n$ be fixed and let $m(b) = max(b - n, -b)$.

Definition 2.11 *The space $M_n^{m(b)}$ is the subspace of M such that each loop in the diagram represents a homology class strictly greater than $m(b)$.*

Evidently, $M_n \subset M_n^{m(b)} \subset M$, and the inclusions are strict.

Remark 2.1 *$R_{(n,b)}$ is even a 1-cocycle in $M_n^{m(b)}$ and not only in M_n. Indeed, going again through the proof of $R_{(n,b)}(m) = 0$ in the lemma we see that it still holds if $\beta = b - n$, $\alpha = b - n$ and $\gamma = -b$ can not happen. This is evidently the case if the homology class of each loop in the diagram is strictly greater than $m(b)$. We will see later that this is already sufficient for $R_{(n,b)}$ to be a 1-cocycle.*

Of course, the analogue statement holds also for $R_{(a,n)}$.

One also easily verifies, that for $R_{(n,0)}$ and $R_{(0,n)}$ the space $M_n^{m(b)}$ can even by replaced by M_n^{-n}, i.e. the marking of each loop is strictly greater than $-n$.

We consider now the 1-cochains which were introduced in Definition 2.5. Let m be a meridian of a positive quadruple crossing. This time we have to consider one more global type of quadruple crossings.

Lemma 2.3 $R(m) = 0$ *for the 1-cochains from Definition 2.5.*

Proof. We give the details of the proof only for $R_{(hm,n-a)}$. The case $R_{(ml,a)}$ is similar and is left to the reader.

As before, the configurations with non-isolated arrows in $R_{(hm,n-a)}$ can occur only for the global types II and IV. The configuration with isolated arrows can occur only for the global types II, IV and V.

Case V:
$-\bar{P}_3$ contributes if and only if $\alpha = 0$, $\gamma = a$ and $\beta = n - a$. Then it cancels out with P_2.

Case IV:
$-P_2$ contributes if and only if $\beta = 0$, $\alpha = n - a$ and $\gamma = a$. But then it cancels out with $\bar{P}_2$.

$\bar{P}_2$ can also contribute if $\beta = 0$, $\alpha = n - a$ and $\gamma = 0$. But then it cancels out with $-P_4$.

P_3 contributes if and only if $\alpha = 0$, $\beta = n - a$ and $\gamma = a$. But then it cancels out with $-\bar{P}_3$.

P_1 contributes (-1) if and only if $\alpha = 0$, $\gamma = 0$ and $\beta = n - a$. But then it cancels out with $\bar{P}_4$.

Case II:
$-P_1$ contributes if and only if $\alpha = 0$, $\gamma = a$ and $\beta = n - a$. But then it cancels out with $-\bar{P}_4$.

$\bar{P}_1$ contributes if and only if $\beta = n - a$, $\alpha = 0$ and $\gamma = 0$. But then it cancels out with $-P_3$.

P_2 contributes if and only if $\beta = 0$, $\gamma = a$ and $\alpha = n - a$. But then it cancels out with $-\bar{P}_2$.

$-P_3$ can also contribute with $\beta = 0$, $\gamma = 0$ and $\alpha = n - a$. But then it cancels out with $\bar{P}_3$. □

We consider now the 1-cochain which was introduced in Definition 2.6.

Lemma 2.4 $R(m) = 0$ *for the 1-cochain from Definition 2.6.*

Proof. The configurations in $R_{(a,a,n)}$ occur for the global types I, II, III, IV and V.

Case IV:

$-P_2$ contributes if and only if $\beta = a$, $\alpha = 0$ and $\gamma = n - a$. But then it cancels out with $\bar{P}_2$.

P_3 contributes if $\alpha = a$, $\beta = \gamma = 0$. But then it cancels out with $\bar{P}_3$.

Case II:

$\bar{P}_1$ could contribute only if $\alpha = a$, $\beta = \gamma = 0$. But its two contributions cancel out in $R_{(a,a,n)}$.

P_4 contributes if and only if $\alpha = a$, $\beta = \gamma = 0$. But then it cancels out with $-\bar{P}_4$.

The first configuration in $-\bar{P}_4$ can never contribute because $\alpha + \gamma = a \neq n$.

Case III:

$\bar{P}_1$ contributes if and only if $\alpha = \gamma = 0$ and $\beta = a$. But the it cancels out with $\bar{P}_2$.

The second configuration in $-P_2$ can never contribute because $\alpha + \beta = a \neq n$.

Case V:

P_3 contributes if and only if $\alpha = a$, $\gamma = 0$ and $\beta = n - a$. But the it cancels out with $-\bar{P}_3$.

Case I:

$-P_1$ could contribute if and only if $\alpha = a$, $\beta = \gamma = 0$. But then its two contributions cancel out in $R_{(a,a,n)}$.

The first configuration in $\bar{P}_1$ can never contribute because $\alpha + \beta = a \neq n$.

$-P_2$ contributes if and only if $\alpha = a$, $\beta = \gamma = 0$. But then it cancels out with $\bar{P}_2$.

P_4 contributes if and only if $\alpha = a$, $\beta = \gamma = 0$. But then it cancels out with $-\bar{P}_4$. $\square$

Remark 2.2 *An important point is that the contribution of P_i to $R_{(a,a,n)}$ cancels always out with the contribution of $\bar{P}_i$, besides for the global type III. Here the contribution of $\bar{P}_1$ cancels out with that of $\bar{P}_2$. Unfortunately, this prevents us to quantize $R_{(a,a,n)}$ in the same way as we will quantize the 1-cocycles for closed braids in Chapter 4. However, the triple crossing $\bar{P}_1$ comes from the local branches $1, 3, 4$ and the triple crossing $\bar{P}_2$ comes from the local branches $2, 3, 4$. Hence, they share exactly the crossing 34 and this crossing corresponds to the crossing hm for both of them. This implies that for each component a_i of the resolution of the trace graph $TG(\gamma)$ we have that $R_{(a,a_i,n)}(\gamma)$ is invariant under generic homotopies of γ through a positive quadruple crossing.*

2.4.3   Cube equations

We have to consider now all other local types of triple crossings together with the self-tangencies. We know already the contributions for the local type 1, i.e. the positive triple crossings, from our solution of the tetrahedron equation and we will determine the contributions of all other local types from the cube equations. The local types of triple crossings were shown in Fig. 2.31. The diagrams which correspond to the edges of the graph Γ (compare Section 2.4.1) are shown in Fig. 2.64. The projection of a triple crossing p into the plan separates the plan near p into three couples of a region and its dual. The regions correspond exactly to the three edges adjacent to the vertex corresponding to the local type of the triple crossing (we can forget about the three dual regions because of $\Sigma^{(3)}_{trans-self-flex}$ as was explained in Section 2.4.1). We show the corresponding graph Γ now in Fig. 2.65. The unfolding of e.g. the edge $1 - 5$ was shown in Fig. 2.36 (compare Chapter 4 and [20]).

Observation 2.1 *The diagrams corresponding to the two vertices of an edge differ just by the two crossings of the self-tangency which replace each other in the triangle as shown in Fig. 2.66. Consequently, the two vertices of an edge have always the same global type and always different local types.*

It follows that we have to solve the cube equations for the graph Γ for each global type of triple crossings. We show the Gauss diagrams for some edges of Γ again in the case M_1 in Fig. 2.67 up to Fig. 2.81. Here, a, b and c are the possibilities for the point at infinity on the long knot.

We represent the meridian m of $\Sigma^{(2)}_{trans-self}$ in the following way: we create first the self-tangency and then we move the transverse branch from the left to the right and we eliminate again the self-tangency.

Lemma 2.5 *Let m be a meridian of $\Sigma^{(2)}_{trans-self}$ or a loop in Γ. Then $R(m) = 0$ for all 1-cochains which were introduced in Section 2.1.*

Proof. We observe from the figures that the arrow of the weights for all 1-cochains from Definitions 2.3 and 2.4 can never become an arrow in the triangle for all strata of $\Sigma^{(2)}_{trans-self}$, no matter the self-tangency has equal or opposite orientation of the two branches. The same is true for $R_{(a,a,n)}$ if the self-tangency has opposite tangent directions. If it has equal tangent direction then we see from the figures, that we have only to consider the

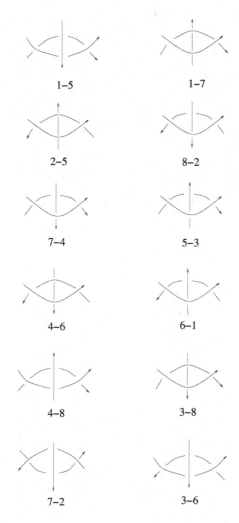

Figure 2.64: The twelve edges of the graph Γ

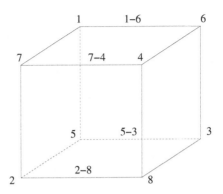

Figure 2.65: The graph Γ

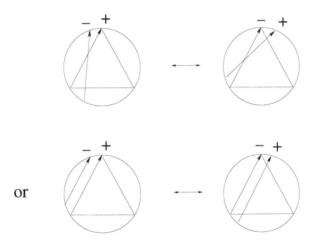

Figure 2.66: Two crossings replace each other for an edge of Γ

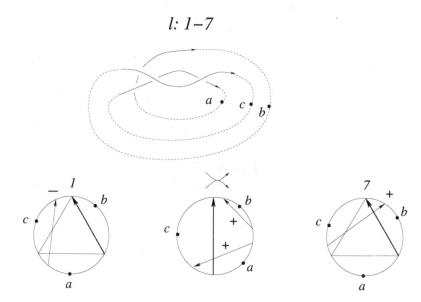

l: 1–7

Figure 2.67: $l1 - 7$

case when ml is a crossing in the self-tangency. The other crossing c of the self-tangency contributes $w(c)$ to $R_{(a,a,n)}$ if it is the arrow of the weight in the second configuration in the definition of $R_{(a,a,n)}$. The crossing c becomes the crossing ml for the other triple crossing in the meridian of $\Sigma^{(2)}_{trans-self}$ and the crossing ml becomes a crossing c'. Evidently, we have $w(c) = -w(c')$. But the crossing c' contributes now $-w(c') = w(c)$ in the third configuration in the definition of $R_{(a,a,n)}$. Consequently, the contributions of the two triple crossings do not change, but their signs are opposite, and they cancel out together. $\square$

2.4.4 Wandering cusps

We have to deal now with the irreducible strata of codimension two which contain a diagram with a cusp which moves over or under another branch. We can assume that in the local picture there is exactly one crossing, before the small curl from the cusp appears. Notice that in this case the local types 2 and 6 can evidently not occur as triple crossings. There are exactly sixteen

$l{:}1{-}5$

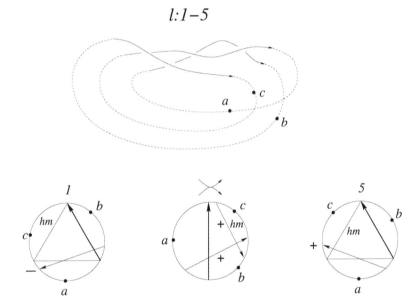

Figure 2.68: $l1 - 5$

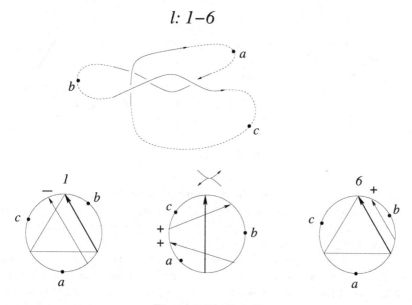

Figure 2.69: $l1 - 6$

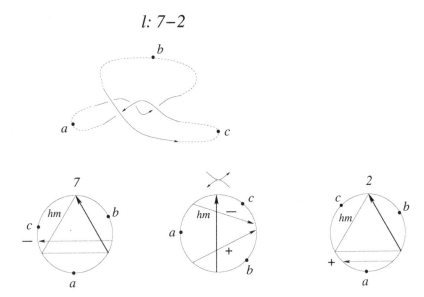

Figure 2.70: $l7 - 2$

possible local types. We list them again for M_1 in Fig. 2.92 to Fig. 2.95, where we move the branch from the right to the left. For each local type we have to consider all global types. Notice, that the new crossing from the cusp has always the marking 0 or n.

We give in the figure also the Gauss diagrams of the triple crossing and of one of the self-tangencies. Notice that in each Gauss diagram of a triple crossing one of the three arcs is empty besides just one head or one foot of an arrow. Let us denote each stratum of $\Sigma^{(2)}_{trans-cusp}$ simply by the global type of the corresponding triple crossing.

The following lemma reduces the number of cases which have to be considered.

Lemma 2.6 *If $R(m) = 0$ for $\Sigma^{(2)}_{trans-cusp}$ with one orientation of the moving branch then it is also 0 for the other orientation of the moving branch.*

Proof. We can change the orientation as shown in Fig. 2.96. This does not change $R(m)$ because R II moves do not contribute at all. □

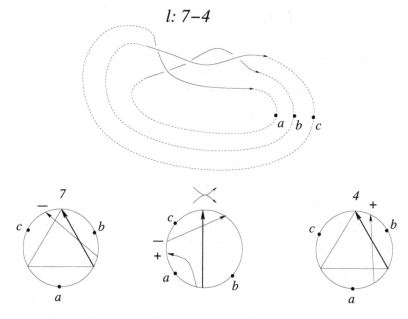

Figure 2.71: $l7 - 4$

l: 3–6

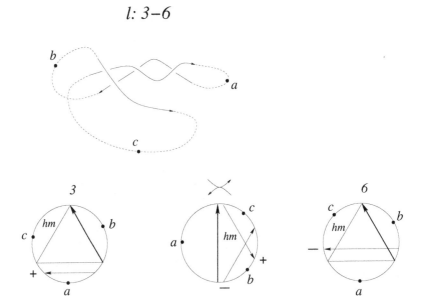

Figure 2.72: $l3 - 6$

$l: 3\text{--}8$

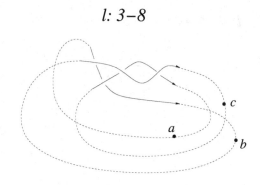

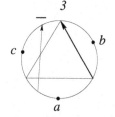
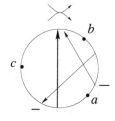
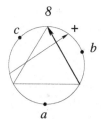

Figure 2.73: $l3 - 8$

l: 4–6

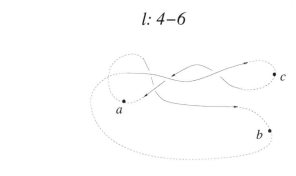

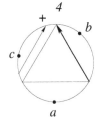
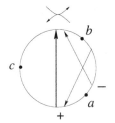
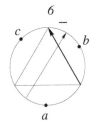

Figure 2.74: l4 − 6

$l: 4-8$

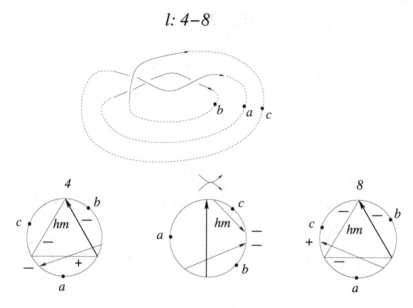

Figure 2.75: $l4 - 8$

$l: 5-2$

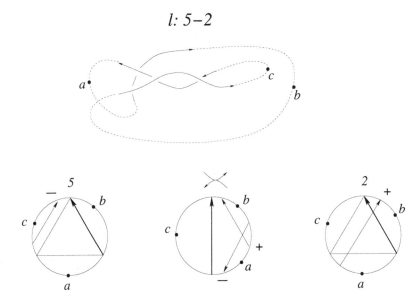

Figure 2.76: $l5 - 2$

l: 5–3

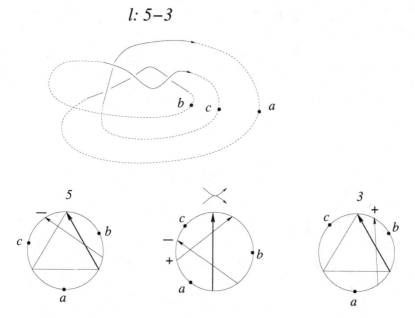

Figure 2.77: $l5 - 3$

l: 8–2

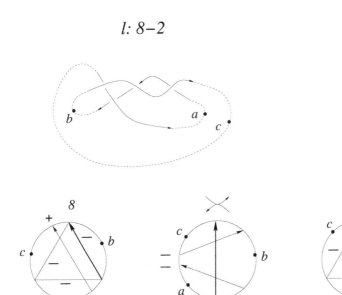

Figure 2.78: l8 − 2

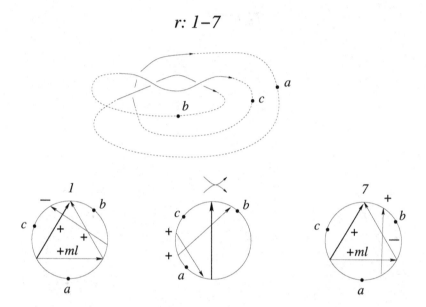

Figure 2.79: $r1 - 7$

r: 1–5

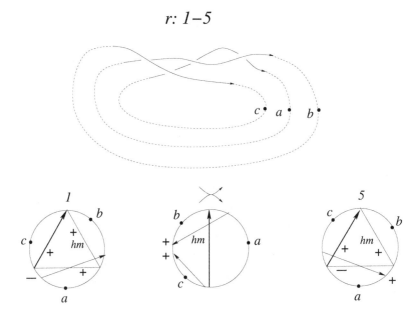

Figure 2.80: $r1 - 5$

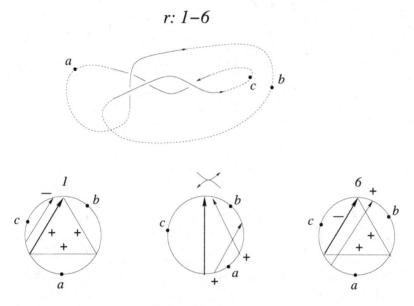

Figure 2.81: $r1 - 6$

r: 3–6

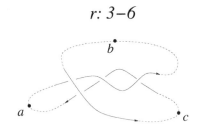

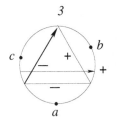
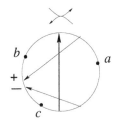
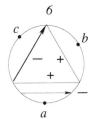

Figure 2.82: $r3 - 6$

r: 8–2

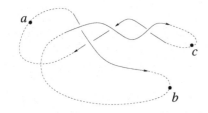

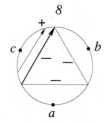
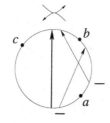
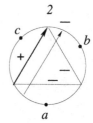

Figure 2.83: $r8 - 2$

r: 4–6

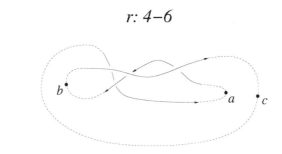

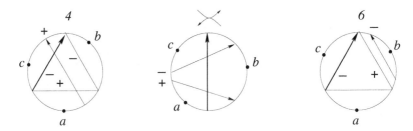

Figure 2.84: $r4 - 6$

r: 3−8

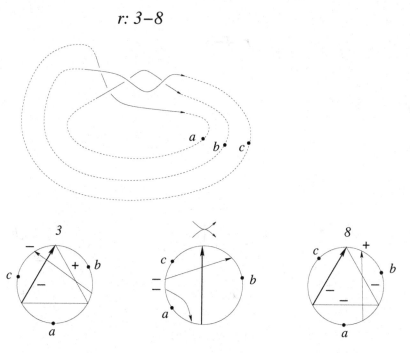

Figure 2.85: $r3 - 8$

r: 4–8

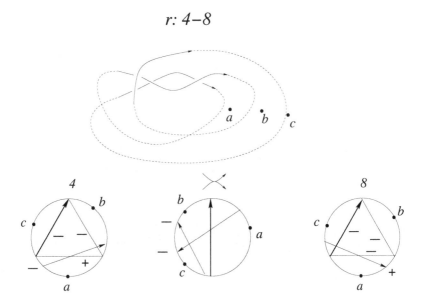

Figure 2.86: $r4 - 8$

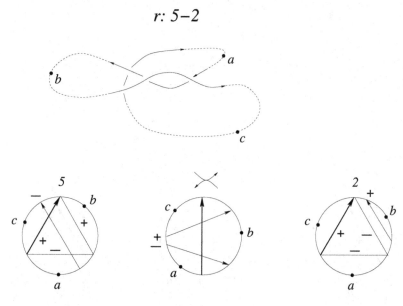

Figure 2.87: $r5 - 2$

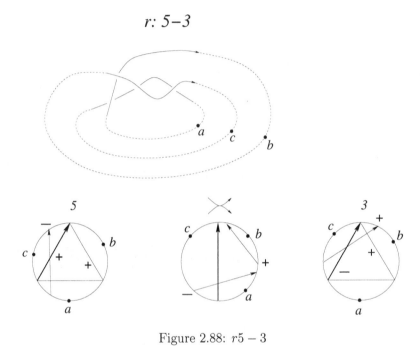

Figure 2.88: $r5 - 3$

r: 7–2

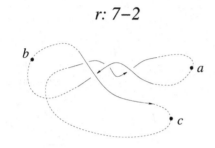

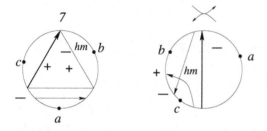

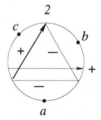

Figure 2.89: $r7 - 2$

r: 7–4

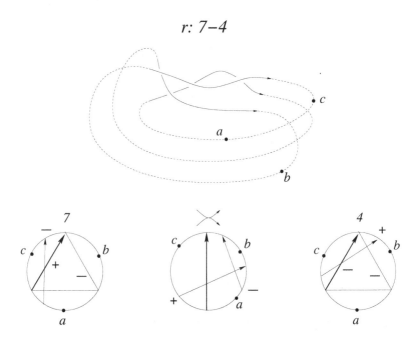

Figure 2.90: $r7 - 4$

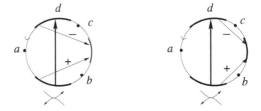

Figure 2.91: The two self-tangencies for the edge $l7 - 2$

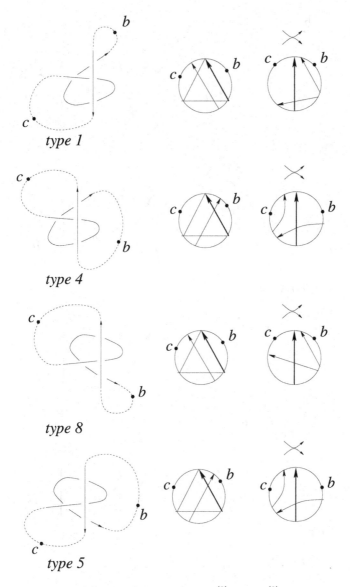

Figure 2.92: The strata $\Sigma^{(2)}_{l_c}$ and $\Sigma^{(2)}_{l_b}$

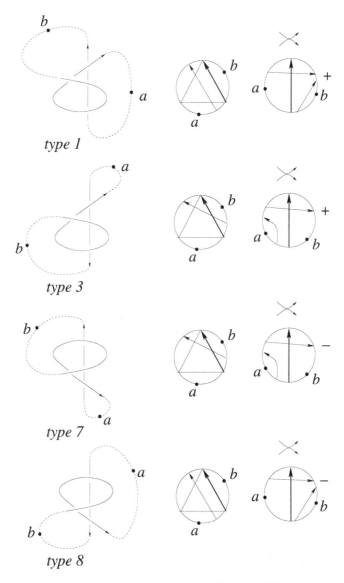

Figure 2.93: The strata $\Sigma_{l_a}^{(2)}$ and $\Sigma_{l_b}^{(2)}$

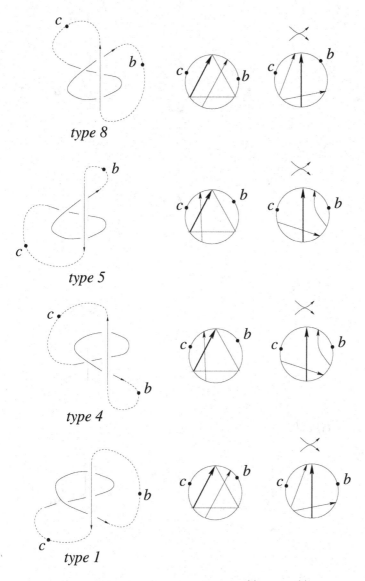

type 8

type 5

type 4

type 1

Figure 2.94: The strata $\Sigma_{r_b}^{(2)}$ and $\Sigma_{r_c}^{(2)}$

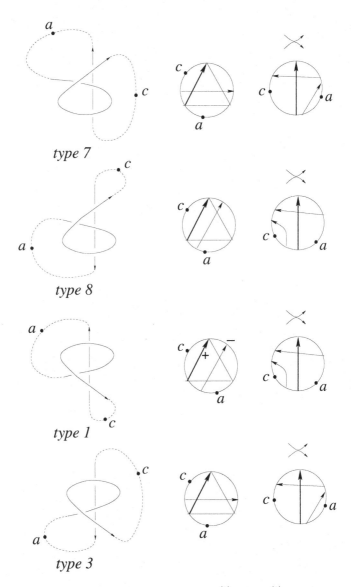

type 7

type 8

type 1

type 3

Figure 2.95: The strata $\Sigma_{r_a}^{(2)}$ and $\Sigma_{r_c}^{(2)}$

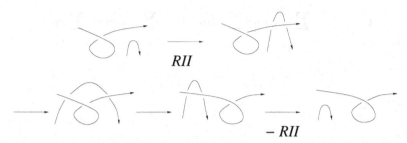

Figure 2.96: Replacing the orientation of the moving branch

Lemma 2.7 *Let m be a meridian of $\Sigma^{(2)}_{trans-cusp}$. Then $R(m) = 0$ for all 1-cochains which were introduced in Section 2.1.*

Proof. The meridian m contains exactly one triple crossing. The new crossing from the cusp has always marking 0 or n and it can never be the crossing d in the triple crossing. Moreover, we see from the figures that the arc in the circle of the Gauss diagram, which corresponds to the small curl which appears from the cusp, is almost empty. There can be only one foot or one heat of an arrow on it, say an arrow c, which corresponds then necessarily to a crossing of a self-tangency in the meridian m. The second crossing of the self-tangency has to correspond then to an arrow in the triangle. It follows immediately that the arrow c can not contribute to any weight of the 1-cochains introduced in Definitions 2.3 and 2.4. Consequently, the triple crossing does not contribute non-trivially to R.

The same is true for the first configuration in the formula for $R_{(a,a,n)}$. The arrows of marking n in the weights of the second and third configurations of $R_{(a,a,n)}$ could a priory be arrows from a self-tangency in the meridian m. We see from the figures that such a configuration from $R_{(a,a,n)}$ appears only in two cases:

In Fig. 2.94, namely for the local type 8 and for the local type 1. However, in both cases the arrow corresponds to the second arrow d in the self-tangency and hence it has marking $a \neq n$.

The other case is shown in Fig. 2.95, again for the local types 8 and 1. This time, in both cases the arrow corresponds to the second arrow ml in the self-tangency. But the crossing from the cusp corresponds now to the crossing hm in the triangle and it has marking 0 or n. Consequently, this R III move is not of the global type (a, a, n) and it does not contribute to $R_{(a,a,n)}$ neither. $\square$

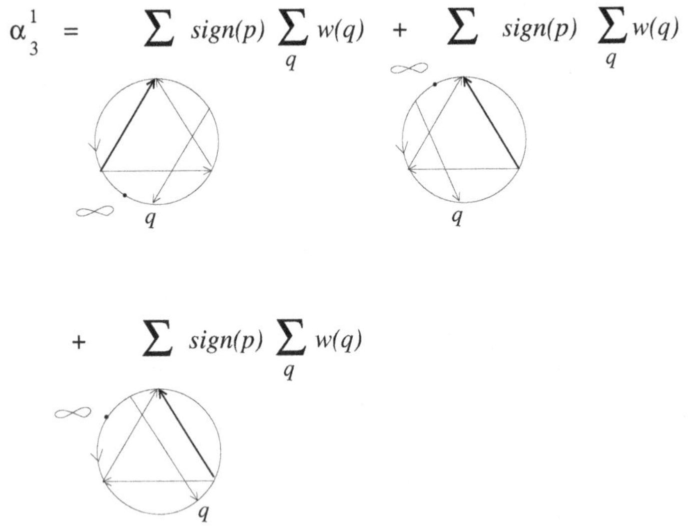

$$\alpha_3^1 \;=\; \sum sign(p) \sum_q w(q) \;+\; \sum sign(p) \sum_q w(q)$$

$$+\; \sum sign(p) \sum_q w(q)$$

Figure 2.97: A formula for Mortier's diagrammatic 1-cocycle α_3^1

2.4.5   Completion of the proofs

We have proven that for all the 1-cochains R introduced in Section 2.1, the value $R(\gamma)$ is invariant under all generic homotopies of γ in M_n through the six types of strata of Proposition 1.1. Consequently, they all define 1-cocycles in M_n. The examples in Section 2.3 show that they are non-trivial (besides for $R_{(a,b)}$, $R_{(a,n-a)}$ and $R_{(n,n)}$) and depend non-trivially on the parameter a.

Notice that for the cube equations and for the wandering cusps we haven't used at all that the knots are in M_n instead of M.

Mortier's integer 1-cocycle α_3^1 for long knots, see [40], [41], is given in Fig. 2.97, where we sum over all Reidemeister III moves p in an oriented loop. His definition of $sign(p)$ is equivalent to our definition (compare Section 2.1). The point on the circle is the point at infinity on the long knot.

We see immediately from the formulas that for long knots α_3^1 coincide with $R_{(0,1)}$ even as a 1-cocycle and not only as a 1-cohomology class.

This finishes the proof of Theorems 2.1, 2.2 and 2.3.

Remark 2.3 *Arnaud Mortier has proven that the reduction modulo 2 of α_3^1 coincides with the reduction modulo 2 of the Teiblum-Turchin 1-cocycle v_3^1 and that $\alpha_3^1(rot(K)) = v_2(K)$. This implies that α_3^1 is in general non-trivial. Mortier has also proven that for framed long knots $\alpha_3^1(fh(K)) = 6v_3(K) - w(K)v_2(K)$, where $v_2(K)$ and $v_3(K)$ are the Vassiliev invariants of order 2 and 3.*

The 1-cochain $R_{(a,a_i,n)}$ is invariant under homotopies through positive quadruple crossings, as was shown in Remark 2.2. By passing with a homotopy of γ a stratum of $\Sigma^{(2)}_{trans-self}$, either hm is the same crossing in the two triple crossings or the two crossings hm are born together in a R II move (compare Section 2.4.3). But then they belong to the same component a_i of the trace graph. Hence $R_{(a,a_i,n)}$ satisfies the cube equations too. If for a wandering cusp the crossing of the cusp is a crossing ml in the R III move p and the crossing hm belongs to a_i then $R_{(a,a_i,n)}(p) = 0$ (compare Section 2.4.4) and $R_{(a,a_i,n)}$ stays invariant. If the crossing of the cusp is hm in the R III move then its marking is not a. Hence, the move does not contribute to $R_{(a,a_i,n)}$. It follows that $R_{(a,a_i,n)}$, and hence also $w(a_i)R_{(a,a_i,n)}$ (compare Definition 2.7), are invariant under all homotopies of γ which do not change the topological type of the resolution of the trace graph $TG(\gamma)$. Consequently, $qR_{(a,a,n)} = \sum_{i=1}^l w(a_i)x^{w(a_i)R_{(a,a_i,n)}}$ is well defined too.

Let K be a knot and γ a loop which contains K. Let us add two new crossings of marking a to K with a R II move and let us consider the corresponding induced new loop γ'. In γ' we have two new components of the trace graph $TG(\gamma')$ with respect to $TG(\gamma)$: a component a_1 and a component a_2, corresponding to the two new crossings from the R II move. Evidently, $R_{(a,a_1,n)}(\gamma') = -R_{(a,a_2,n)}(\gamma')$ and $w(a_1) = -w(a_2)$. It follows that $w(a_1)x^{w(a_1)R_{(a,a_1,n)}(\gamma')} = -w(a_2)x^{w(a_2)R_{(a,a_2,n)}(\gamma')}$ and $qR_{(a,a,n)}(\gamma')$ is still well-defined.

A Morse modification of index 0 or 2 of $TG(\gamma)$ adds or eliminates a component a_i with $w(a_i) = 0$ and hence does not change $qR_{(a,a,n)}(\gamma)$. Morse modifications of index 1 correspond exactly to tangencies of γ from the positive side with strata from $\Sigma^{(1)}_{tan}$. But they are excluded in Proposition 2.1.

This finishes the proof of Proposition 2.1.

3

A 1-cocycle for all knots and all loops in the solid torus

We consider now M instead of M_n.

Let $n \neq 1$. Instead of the homological markings $a \in \mathbb{Z}$ we consider the induced homological markings $\bar{a} \in \mathbb{Z}/n\mathbb{Z}$ (hence, for $n = 0$ the markings $a = \bar{a}$ stay in $\mathbb{Z}$). Surprisingly, the 1-cocycle defined in Definition 2.6 carries over with the new markings to a 1-cocycle in M, in contrast to the 1-cocycles introduced in the Definitions 2.3 and 2.5.

Definition 3.1 *Let $n \neq 1$ and $\bar{a} \neq \bar{0}$.*
The 1-cochain $R_{\bar{a}}$ in M is defined in Fig. 3.1.

Theorem 3.1 *The 1-cochain $R_{\bar{a}}$ is a 1-cocycle in M for each knot K and each $\bar{a} \neq 0$. It represents in general a non-trivial cohomology class in $H^1(M; \mathbb{Z})$.*

Proof. Going again through the proof of Lemma 2.4 we notice that all the same arguments apply when we replace a by $\bar{a}$ and 0 and n by $\bar{0}$, because

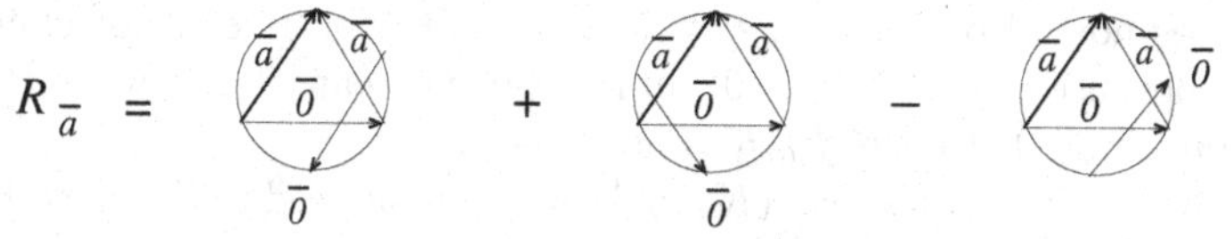

Figure 3.1: The 1-cocycle $R_{\bar{a}}$

111

still $\bar{a} \neq \bar{0}$. We observe, that whenever we need that e.g. $(\alpha + \gamma) = \bar{0}$, then we need already that $\bar{\alpha} = \bar{0}$ or $\bar{\gamma} = \bar{0}$. Consequently, $(\alpha + \gamma) = \bar{0}$ implies again $\bar{\alpha} = \bar{0}$ and $\bar{\gamma} = \bar{0}$, analogue to the case of $R_{(a,a,n)}$.

Going again through the proofs of Lemma 2.5 and Lemma 2.7, we see that exactly the same arguments still apply because we had not used here at all that the knot is in M_n instead of M.

This finishes the proof of Theorem 3.1. □

Remark 3.1 *Notice that* $R_{\bar{a}}(\gamma) = R_{(a,a,n)}(\gamma)$ *for each loop* $\gamma \subset M_n$, *if* $n > 1$. *Indeed, if we replace* $0 < a < n$ *by* $a+n$ *or* $a-n$ *then we would have negative loops in the diagram. If there are no negative loops then* $\bar{0}$ *in the triangle has to be* n *and the arrow* $\bar{0}$ *in the first configuration has the marking* 0. *The arrows* $\bar{0}$ *in the two remaining configurations have necessarily the marking* n. *Consequently, the formula for* $R_{\bar{a}}$ *reduces just to the formula of* $R_{(a,a,n)}$ *if* $\gamma \subset M_n$. *Consequently, e.g. the Examples 2.3 and 2.4 show immediately that* $R_{\bar{a}}(\gamma) \neq 0$ *too. But this time we can even conclude that* $[\gamma] \neq 0$ *in* $H^1(M)$ *and not only in* $H^1(M_n)$.

If $n \neq 0$ then the polynomial $LR_{\bar{a}}$ is defined exactly as $LR_{(a,a,n)}$ was defined previously. Evidently, its degree $deg(LR_{\bar{a}}) \leq n - 2$. If $n = 0$ then we take the Lagrange interpolation polynomial for all values between the greatest and the smallest a such that the value of $R_{\bar{a}}$ is non-trivial. This polynomial can have an arbitrarily high degree for a fixed n.

$R_{\bar{a}}$ can be quantized in the free loop space $\mathbb{L}M$ exactly in the same way as $R_{(a,a,n)}$ was quantized in $\mathbb{L}M_n$.

It follows directly from the formula that $R_{\bar{a}}$ is automatically trivial if the knot is isotopic to a knot contained in a 3-ball in V (hence there are no crossings of marking $\bar{a} \neq \bar{0}$) and it is also automatically trivial if the knot is isotopic to a closed braid in V (hence there are no crossings of marking $\bar{0}$).

It is amazing that our other 1-cocycles can *not* be generalized to 1-cocycles in M by reducing the markings to markings in $\mathbb{Z}/n\mathbb{Z}$. Indeed, let us consider e.g. $R_{(n,b)}$. In *Case IV* of the positive global tetrahedron equation: $-\bar{P}_3$ contributes now also if $\alpha = \bar{b}$, $\gamma = (-\bar{b})$ and $\beta = \bar{b}$. It can not cancel out with P_2 because $\alpha + \beta + \gamma \neq \bar{0}$, and it can not cancel out with P_1 or P_4 because $n - \gamma \neq \bar{0}$. Hence, $R_{(\bar{0},\bar{b})}$ is not a 1-cocycle.

The loops $rot(K)$ and $slide(K)$ in M were introduced in the Introduction. In fact, it is not very difficult to prove that these loops represent in general non-trivial homology classes in $H_1(M)$ by using the homology classes of the

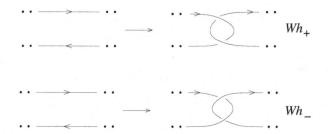

Figure 3.2: The two Whitehead doubles of a framed knot

components of the resolution of the trace graphs TG for these loops, compare Chapter 4 and [20], because Morse modifications change the homotopy type of TG but not the homology class represented by TG.

Let K be an oriented framed knot in V.

Definition 3.2 *The* positive (negative) Whitehead double *of $K \subset V$, denoted by $Wh_+(K)$ respectively $Wh_-(K)$, is the knot which is obtained by two parallel copies of K, which are oriented differently and are closed two a knot by the modifications shown in Fig. 3.2.*

Evidently, $Wh_+(K)$ and $Wh_-(K)$ are 0-homologous in the solid torus for all framed knots K.

Let K' be a framed oriented knot in S^3. First we transform K' into a long knot and replace the long knot by its untwisted parallel n-cable, possibly with changing the orientation of some of the components. We close now the n-cable with an appropriate fixed oriented n-component string link T to obtain an oriented knot K in the standard solid torus V. For example, we could consider Whitehead doubles of Whitehead doubles and so on. An example of a closure T for a 4-cable is shown in Fig. 3.3.

We can apply now our 1-cocycle $R_{\bar{a}}$ to several loops in M: pushing T once along the parallel n-cable of K'; the loop induced by the rotation of V around its core; the Fox-Hatcher loop applied only to the n-cable of K' with a fixed position of T. At the end we push T back into the initial position. We illustrate the last loop in Fig. 3.4, where the last two arrows correspond to pushing T along the knot. If we have n strands then we have of course to move the bunch of the n strands over or under the rest of the knot and perhaps to push full-twists of the n strands back through the knot.

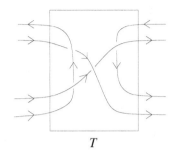

Figure 3.3: A possible closure T of a parallel 4-cable in M

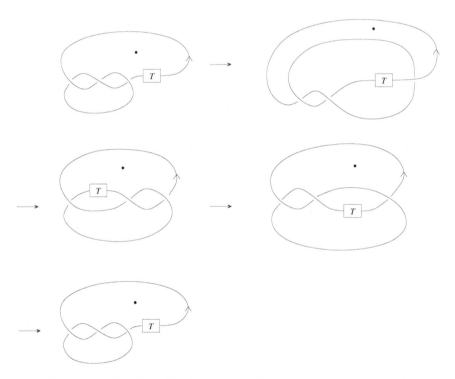

Figure 3.4: The Fox-Hatcher loop in the presence of a string link

The values of $R_{\bar{a}}$ on these loops are not only invariants of $K \subset V$ but also of $K' \subset S^3$, simply because framed knots in S^3 are regularly isotopic (i.e. without Reidemeister I moves) if and only if the corresponding long knots are regularly isotopic as long knots (see e.g. [16]). The interesting point is, that the weights of the triple crossings in the 1-cocycle mix contributions from crossings of T with contributions of crossings of K' for the same triple crossing, as was shown in Example 2.4.

It could also be very interesting to study the *bilinear forms* on the free $\mathbb{Z}$-module generated by all framed knot types of classical knots. Given two framed classical knots K' and K'' and a natural number n, we can take T as the parallel n-cable of K'' twisted with the standard permutation braid $\sigma_1 \sigma_2 \ldots \sigma_{n-1}$, we denote this T still by K'', and we push it once in the solid torus through the parallel n-cable of K'. The value of $LR_{\bar{a}}(push(K'', K'))$ on this loop is an invariant of the ordered couple (K'', K').

This invariant is not always trivial already for $n = 2$. Let K'' be the trivial knot with trivial framing and let K' be the standard diagram of the positive trefoil. Then evidently $K'' = \sigma_1$ and a calculation analogous to that of Example 2.3 gives $LR_{\bar{a}}(push(K'', K')) = R_{\bar{1}}(push(K'', K')) = -1$.

But unfortunately it is too hard to calculate more interesting examples by hand.

4

Polynomial 1-cocycles for closed braids in the solid torus

We start with developing 1-parameter knot theory in general, following [20]. People who are more interested in braids than in knots can start reading from Section 4.7.

4.1 Introduction to 1-parameter knot theory

4.1.1 Motivation and summary

The classical Reidemeister theorem says that plane diagrams represent isotopic links in 3-space if and only if they can be related by finitely many moves of 3 types corresponding to the codimension 1 singularities of links diagrams, namely a triple point ✕, simple tangency ✕ and ordinary cusp ✗.

We establish the higher order Reidemeister theorem considering a canonical 1-parameter family of links in a solid torus and studying codimension 2 singularities of resulting link diagrams. The 1-parameter family of link diagrams is encoded by a new combinatorial object, a trace graph in a thickened torus in such a way that *trace graphs determine families of isotopic links if and only if they can be related by a finite sequence of the 11 moves in Fig. 4.11*, see Theorem 4.1.

The conjugacy problem for braids is equivalent to the isotopy classification of closed braids in a solid torus. Braids are *conjugate* if and only if the trace graphs of their closures are *equivalent* through only tetrahedral moves

117

and trihedral moves in Figs. 4.11i, 4.11xi. Trace graphs of closed braids can be recognized up to isotopy in a thickened torus and trihedral moves in polynomial time with respect to the braid length, see Theorem 4.2. The method provides a new geometric approach to the conjugacy problem for braid groups B_n, which still has no efficient solution for $n \geq 5$ strands, i.e. with a polynomial complexity in the braid length. Very promising steps towards a polynomial solution were made by Birman, Gebhardt, González-Meneses [7] and Ko, Lee [37]. A clear obstruction is that the number of different conjugacy classes of braids grows exponentially even in B_3, see Murasugi [46].

Usually links are studied in terms of braids using the theorems of Alexander and Markov, see Birman [5]. The 1-parameter approach is a geometric alternative to the algebraic one: conjugacy of braids and Markov moves are replaced by a stronger notion of link isotopy and extreme tangency moves in Fig. 4.11viii, respectively.

4.1.2 Basic definitions

We work in the C^∞-smooth category.

Fix Euclidean coordinates x, y, z in $\mathbb{R}^3$.

Denote by D_{xy} the unit disk with center at the origin of the horizontal plane XY.

Introduce the solid torus $V = D_{xy} \times S^1_z$, where the oriented circle S^1_z is the segment $[-1, 1]_z$ with the identified endpoints, see the left picture of Fig. 4.1.

Definition 4.1 *An* embedding *is a diffeomorphism onto its image. An oriented link* $K \subset V$ *is the image of an embedding* $f : \sqcup_{j=1}^m S^1_j \to V$. *An* isotopy *between two oriented links* K_0 *and* K_1 *in* V *is a smooth map* $F : (\sqcup_{j=1}^m S^1_j) \times [0, 1] \to V$ *such that* $f_0(\sqcup_{j=1}^m S^1_j) = K_0$, $f_1(\sqcup_{j=1}^m S^1_j) = K_1$ *and the maps* $f_r = F(*, r) : \sqcup_{j=1}^m S^1_j \to V$ *are smooth embeddings for all* $r \in [0, 1]$.

Mark n *points* $p_1, \dots, p_n \in D_{xy}$. *A braid* β *on* n *strands is the image of a smooth embedding of* n *segments into* $D_{xy} \times [-1, 1]_z$ *such that (see Fig. 4.1)*

- *the strands of* β *are monotonic with respect to* $\mathrm{pr}_z : \beta \to S^1_z$;

- *the lower and upper endpoints of* β *are* $\cup(p_i \times \{-1\})$, $\cup(p_i \times \{1\})$, *respectively.*

Braids are considered up to isotopy in the cylinder $D_{xy} \times [-1, 1]_z$, *fixed on its boundary. The isotopy classes of braids form the group denoted by* B_n.

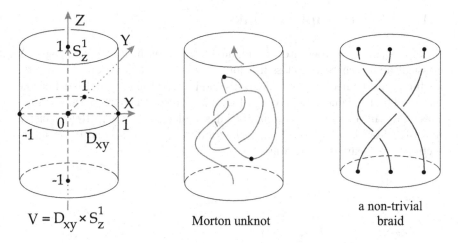

Figure 4.1: Basic notations and examples

The trivial braid consists of n vertical segments $\sqcup_{i=1}^{n}(p_i \times [-1,1]_z)$. A braid $\beta \in B_n$ is pure if the induced permutation $\tilde{\beta} \in S_n$ is its endpoints is trivial. The closed braid $\hat{\beta} \subset V$ is obtained from $\beta \subset D_{xy} \times [-1,1]_z$ by identifying the bases $\{z = \pm 1\}$. ∎

The smoothness of a link K implies that the tangent vector $\dot{f}(s)$ never vanishes on K. The *standard* unknot is given by the trivial embedding $S_z^1 \to V = D_{xy} \times S_z^1$. We introduce a new equivalence relation, *strong isotopy*, for links in a solid torus. For closed braids, the usual isotopy through closed braids is strong.

Definition 4.2 *An* extreme pair *of a link $K \subset V$ is a pair of either 2 local maxima $\wedge\wedge$ or 2 local minima $\vee\vee$ of the projection $\mathrm{pr}_z : K \to S_z^1$ with the same z-coordinate. A smooth isotopy $F : (\sqcup_{j=1}^{m} S_j^1) \times [0,1] \to V$ of links is called* strong *if all links in the family $K_r = F(\sqcup_{j=1}^{m} S_j^1, r) \subset V$ have* no extreme pairs *for $r \in [0,1]$.* ∎

H. Morton proposed the trivial knot in the middle picture of Fig. 4.1. The *Morton* unknot is not strongly isotopic to the *standard* unknot S_z^1. The arc between the marked extrema is a long trefoil that can not be unknotted by strong isotopy since the marked extrema remain the highest and lowest critical points.

4.1.3   Trace graphs of links

Links are usually represented by plane diagrams with double crossings. A classical approach to the classification of links is to use isotopy invariants, i.e. functions defined on plane diagrams and invariant under the Reidemeister moves. The Reidemeister moves in Fig. 4.5 correspond to simplest singularities that can appear in diagrams of links under isotopy, e.g. Reidemeister move III describes the change of a diagram when a transversal triple intersection $\times\!\!\!\times$ appears in the projection.

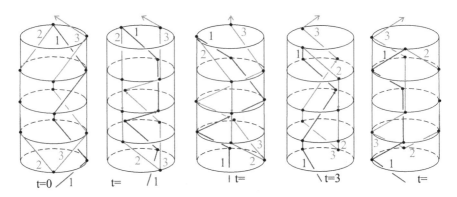

Figure 4.2: Diagrams of rotated trefoils $\mathrm{rot}_t(K) \subset V$ for $t \in [0, \pi]$

The analogue of a plane diagram in the 1-parameter approach is a 1-parameter family of diagrams obtained by rotating a link in V around S_z^1. This is a 2-dimensional surface containing more information about the link than only one plane diagram. The link will be reconstructed up to smooth equivalence from the self-intersection of the surface, the *trace graph*. Denote by $\mathrm{rot}_t : V \to V$ the *rotation* of the torus V around S_z^1 through an angle $t \in [0, 2\pi)$, see Fig. 4.2. Here t is the parameter on *the time circle* $S_t^1 = \mathbb{R}/2\pi\mathbb{Z}$ of length 2π. Let A_{xz} be the vertical annulus $[-1, 1]_x \times S_z^1$ in the solid torus V. Define the *thickened torus* $\mathbb{T} = \mathrm{A}_{xz} \times S_t^1$. We illustrate the rotation of V using the piecewise linear trefoil $K \subset V$ in Fig. 4.2, which can be easily smoothed. Diagrams of rotated trefoils $\mathrm{rot}_t(K) \subset V$ under the orthogonal *projection* $\mathrm{pr}_{xz} : V \to \mathrm{A}_{xz}$ are shown in Fig. 4.2.

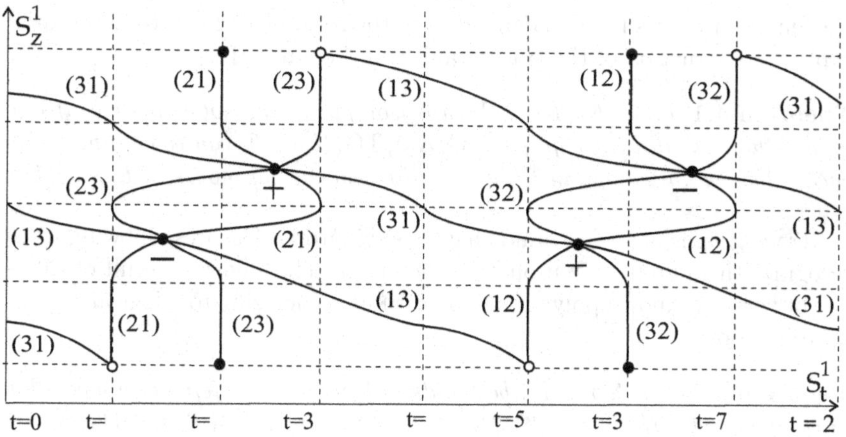

Figure 4.3: The trace graph TG(K) obtained from the diagrams in Fig. 4.2

Definition 4.3 *The* trace graph $\mathrm{TG}(K) \subset \mathbb{T}$ *of a link* $K \subset V$ *consists of the crossings of the diagrams* $\mathrm{pr}_{xz}(\mathrm{rot}_t(K))$ *over all* $t \in S^1_t$. *Mark the intersection points from* $K \cap (\mathrm{D}_{xy} \times \{\pm 1\}) \subset V$ *and also mark each local extremum of* K *with respect to* $\mathrm{pr}_z : K \to S^1_z$. *If* K *has* m *components, in general position, the* i*-th component decomposes into* n_i *vertically monotonic arcs labelled by* A_{iq}, $q = 1, \ldots, n_i$. *The 3 monotonic arcs in Fig. 4.2 are numbered simply by 1, 2, 3.*

Take a point $p \in \mathrm{TG}(K)$, *which is a crossing of* A_{iq} *over* A_{js} *in the diagram* $\mathrm{pr}_{xz}(\mathrm{rot}_t(K))$ *for some* $t \in S^1_t$. *Associate to* p *the ordered* label $(q_i s_j)$. *Then the edges of the graph* $\mathrm{TG}(K)$ *are labelled. In the case* $m = 1$ *we miss the indices* i, j *of components of* K *and label edges by ordered pairs* (qs), *see Fig. 4.3.* ∎

For each $t \in S^1_t$, watch the crossings of the diagram $\mathrm{pr}_{xz}(\mathrm{rot}_t(K)) \subset \mathrm{A}_{xz} \times \{t\}$, e.g. the initial diagram $\mathrm{pr}_{xz}(K) \subset \mathrm{A}_{xz}$ at $t = 0$ has 3 double crossings, which evolve during the rotation of K. At $t = \pi/4$, the lowest crossing becomes a critical crossing Ψ corresponding to a critical vertex -o- of TG(K). At the same $t = \pi/4$ a couple of crossings is born after Reidemeister move II associated to a tangency $\asymp$. At $t = \pi/2$ a new crossing is born from a cusp Υ after Reidemeister move I, which leads to a hanging vertex $\bullet\!-$ of TG(K). The 2 triple vertices of TG(K) for $t \in (0, \pi)$ correspond to 2

Reidemeister moves III happening during the rotation of K. A combinatorial explicit construction of the trace graph is in Lemma 4.18.

Theorem 4.1 *Links $K_0, K_1 \subset V$ are isotopic in the solid torus V if and only if their labelled trace graphs $\mathrm{TG}(K_0), \mathrm{TG}(K_1) \subset \mathbb{T}$ can be obtained from each other by an isotopy in $\mathbb{T}$ and a finite sequence of moves in Fig. 4.11.*

Trace graphs of closed braids have combinatorial features, allowing us to recognize them up to all but one type of moves. The following result of [20] is one of very few known polynomial algorithms recognizing topological objects up to isotopy.

Theorem 4.2 *Let $\beta, \beta' \in B_n$ be braids of length $\leq l$. There is an algorithm of complexity $C(n/2)^{n^2/8}(6l)^{n^2-n+1}$ to decide whether $\mathrm{TG}(\hat{\beta})$ and $\mathrm{TG}(\hat{\beta'})$ are related by isotopy in $\mathbb{T}$ and trihedral moves, the constant C does not depend on l and n. In the case of pure braids, the power $n^2/8$ can be replaced by 1. If the closure of a braid is a single curve in the solid torus, then the complexity reduces to $Cn(6l)^{n-1}$.*

4.1.4   Scheme of proofs

The plan of the proof for Theorem 4.1 is shown in Fig. 4.4

The *first* double arrow in Fig. 4.4 is a classical reduction of an equivalence of links to extended Reidemeister moves on plane diagrams in Fig. 4.5.

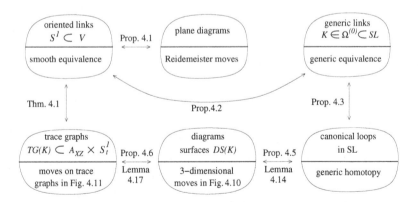

Figure 4.4: A scheme to prove Theorem 4.1

The *second* arrow is a new reduction to generic links and generic equivalence defined in terms of codimension 1 singularities with respect to the rotation of links in V.

The *third* arrow is a reformulation of the previous reduction in terms of canonical loops of links in the space of all links in the solid torus V.

The *fourth* arrow is a reduction of generic links to their 2-dimensional diagram surfaces considered up to 3-dimensional moves in Fig. 4.10.

The *fifth arrow* is a final reduction of generic links to their trace graphs considered up to equivalence generated by the moves in Fig. 4.11.

The key ingredient of the proofs is a description of versal deformations and bifurcation diagrams of codimension 2 multi local singularities of plane curves in Section 4.4.

4.2 Singular subspaces in the space of all links

4.2.1 Codimension 1 singularities of link diagrams

Let M, N be smooth finite dimensional manifolds. Denote by $J_{[l]}^k(M, N)$ the space of all l-tuple k-jets of smooth maps $\xi : M \to N$ for all tuples $(u_1, \ldots, u_l) \in M^l$, see [1, sections I.2]. Let $(x_1, \ldots, x_m)$ and $(y_1, \ldots, y_n)$ be local coordinates in M and N, respectively. If the map ξ is defined locally by $y_j = \xi_j(x_1, \ldots, x_n)$, $j = 1, \ldots, m$, then the l-tuple k-*jet* of the map ξ at a point $(u_1, \ldots, u_l)$ is determined by

$$\{x_1, \ldots, x_m\}; \ \{y_1, \ldots, y_n\}; \ \left\{\frac{\partial \xi_j}{\partial x_i}\right\}; \ \ldots \ \left\{\frac{\partial^k \xi_j}{\partial x_{i_1} \ldots \partial x_{i_s}}\right\}, i_1 + \cdots + i_s = k.$$

The above quantities define local coordinates in $J_{[l]}^k(M, N)$. The l-tuple k-*jet* $j_{[l]}^k \xi$ of a smooth map $\xi : M \to N$ can be considered as the map $j_{[l]}^k \xi : M^l \to J_{[l]}^k(M, N)$, namely $(u_1, \ldots, u_l)$ goes to the l-tuple k-jet of the map ξ at $(u_1, \ldots, u_l)$.

Take an open set $W \subset J_{[l]}^k(M, N)$ for some k, l. The set of smooth maps $f : M \to N$ with l-tuple k-jets from W is called *open*. These sets for all open $W \subset J_{[l]}^k(M, N)$ form a basis of the *Whitney* topology in $C^\infty(M, N)$. So two maps are close in the Whitney topology if they are close with all theirs derivatives.

Definition 4.4 *The* space SL *of all links* $K \subset V$ *inherits the* Whitney *topology from* $C^\infty(\sqcup_{j=1}^m S_j^1, V)$. *A link* K *defined by a smooth embedding* $f : \sqcup_{j=1}^m S_j^1 \to V$ *is called* general *if the diagram* $D = \mathrm{pr}_{xz}(K) \subset A_{xz}$ *is general, namely*

- *the map* $\mathrm{pr}_{xz} \circ f : \sqcup_{j=1}^m S_j^1 \to A_{xz}$ *is a smooth embedding outside finitely many double* crossings, *an overcrossing arc is specified at each crossing;*

- *the extrema of* $\mathrm{pr}_z : D \to S_z^1$ *are not crossings and have distinct z-coordinates.*

Denote by $\Sigma^{(0)} \subset$ SL *the subspace of all general links.*   ∎

We consider local singularities, so fix coordinates x, z around each point in A_{xz}. The x-axis is said to be *horizontal*, i.e. it is perpendicular to the vertical core $S_z^1 \subset A_{xz}$. Classical codimension 1 singularities of plane curves were described by David [15, List I on p. 561], namely the ordinary cusp $\curlyvee$ (the A_2 singularity in Arnold's notations), simple tangency X (A_3) and triple point X (D_4). The solid torus V has the distinguished vertical direction along S_z^1, so we also consider singularities with respect to $\mathrm{pr}_z : V \to S_z^1$, e.g. Reidemeister move IV is generated by passing through a critical crossing $\curlywedge$, where one of the tangents is horizontal.

Definition 4.5 *A diagram D is the image of a smooth map $g : \sqcup_{j=1}^m S_j^1 \to A_{xz}$.*

X *A* triple point *of the diagram D is a transversal intersection p of 3 arcs such that all the tangents at p are not horizontal.*

X *A* simple tangency *is an intersection p of 2 arcs given locally by $u = \pm v^2$. We assume that the tangent at p is not horizontal.*

$\curlyvee$ *An* ordinary cusp *is the singular point p of an arc given locally by $u^2 = v^3$. We assume that the tangent at p is not horizontal.*

$\curlywedge$ *A* critical crossing *is a transversal intersection p of 2 arcs such that one of the tangents at p is horizontal.*

$\curvearrowright$ *A* cubical point *is the singular point p of an arc given locally by $z = u^3$, the tangent at p is horizontal.*

$\wedge^{\vee}$ *A* mixed pair *is a pair of a local maximum and a local minimum of the projection* $\mathrm{pr}_z : D \to S_z^1$, *lying in the same horizontal line.*

$\wedge\wedge$ *An* extreme pair *is a pair of either 2 local maxima or 2 local minima of the projection* $\mathrm{pr}_z : D \to S_z^1$, *lying in the same horizontal line.*

Given a singularity $\gamma \in \{ \mathbb{X}, \mathsf{X}, \Upsilon, \wedge, \diagup, \wedge^{\vee}, \wedge\wedge \}$, *denote by* $\Sigma_\gamma \subset \mathrm{SL}$ *the singular subspace consisting of all links* $K \subset V$ *such that* $\mathrm{pr}_{xz}(K)$ *is general outside* γ.

Set $\qquad \Sigma^{(1)} = \Sigma_{\mathbb{X}} \cup \Sigma_{\mathsf{X}} \cup \Sigma_{\Upsilon} \cup \Sigma_{\wedge} \cup \Sigma_{\diagup} \cup \Sigma_{\wedge^{\vee}} \cup \Sigma_{\wedge\wedge} \subset \mathrm{SL}.$ ∎

4.2.2 Extended Reidemeister theorem

Definition 4.6 *Let* M *be a finite dimensional smooth manifold. A subspace* $\Lambda \subset M$ *is called a* stratified space *if* Λ *is the union of disjoint smooth submanifolds* Λ^i *(*strata*) such that the boundary of each stratum is a finite union of strata of less dimensions. Let* N *be a finite dimensional manifold. A smooth map* $\xi : M \to N$ *is* transversal *to a smooth submanifold* $U \subset N$ *if the spaces* $f_*(T_x M)$ *and* $T_{f(x)}U$ *generate* $T_{f(x)}N$ *for each* $x \in M$. *A smooth map is* $\eta : M \to N$ transversal *to a stratified space* $\Lambda \subset N$ *if the the map* η *is transversal to each stratum of* Λ. ∎

Briefly Theorem 4.3 says that any map can be approximated by 'a nice map'.

Theorem 4.3 (Multi-jet transversality theorem of Thom), *see [1, Section I.2]. Let* M, N *be compact smooth manifolds,* $\Lambda \subset J_{[l]}^k(M, N)$ *be a stratified space. Given a smooth map* $\xi : M \to N$ *there is a smooth map* $\eta : M \to N$ *such that*

- *the map* η *is arbitrarily close to* ξ *with respect to the Whitney topology;*

- *the l-tuple k-jet* $j_{[l]}^k \eta : M^l \to J_{[l]}^k(M, N)$ *is transversal to* $\Lambda \subset J_{[l]}^k(M, N)$.

Lemma 4.1 (a) *The subspace* $\Sigma^{(1)}$ *has codimension 1 in the space* SL.
(b) *The subspace* $\Sigma^{(0)}$ *is open and dense in the space* SL.

Sketch 4.1 (a) *It is a standard computation in the space $J^1_{[3]}(\mathbb{R}, \mathbb{R}^2)$ of 3-tuple 1-jets of maps $(x(r), z(r)) : \mathbb{R} \to \mathbb{R}^2$. For instance, fixing 3 parameters r_1, r_2, r_3, the subspace $\Sigma_{\times}$ maps to the subspace of $J^1_{[3]}(\mathbb{R}, \mathbb{R}^2)$ given by 4 equations $x(r_1) = x(r_2) = x(r_3)$, $z(r_1) = z(r_2) = z(r_3)$ and 3 inequalities $\dot{z}(r_i) \neq 0$, $i = 1, 2, 3$, meaning that the tangents are not horizontal, hence the codimension of the subspace $\Sigma_{\times} \subset \mathrm{SL}$ is 1 after forgetting the 3 parameters. Analogously $\Sigma_{\curlyvee}$ maps to the subspace given by 4 equations $\dot{x}(r_1) = \dot{z}(r_1) = 0$, $r_1 = r_2 = r_3$, hence the codimension of $\Sigma_{\curlyvee} \subset \mathrm{SL}$ is 1. A similar detailed argument will be given in the proof of Lemma 4.3.*

(b) *The conditions of Definition 4.4 define an open subset of SL whose complement is clearly the closure of the codimension 1 subspace $\Sigma^{(1)}$ from Definition 4.5.* ∎

The following result immediately follows from Lemma 4.1 since by Theorem 4.3 any isotopy in the space SL of links can be approximated by a path transversally intersecting the singular subspace $\Sigma^{(1)} \subset \mathrm{SL}$ of codimension 1.

Proposition 4.1 *Any smooth link can be approximated by a general link. General links are isotopic if and only if their diagrams can be obtained from each other by a plane isotopy and finitely many Reidemeister moves in Fig. 4.5.*

In Fig. 4.5 orientations of arcs and symmetric images of the moves are omitted.

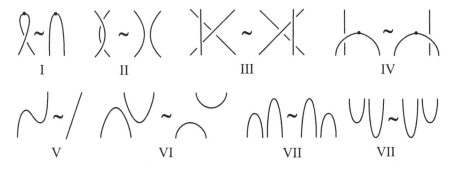

I   II   III   IV

V   VI   VII   VII

Figure 4.5: Reidemeister moves taking into account local extrema

4.2.3 The co-orientation of codimension 1 subspaces

Using Gauss diagrams of link diagrams, we define the co-orientation of codimension 1 subspaces $\Sigma_{\ast}, \Sigma_{\times}, \Sigma_{\curlyvee}, \Sigma_{\curlywedge}$ from Definition 4.5.

Definition 4.7 *Let a general link K be defined by $f : \sqcup_{j=1}^{m} S_j^1 \to V$. The Gauss diagram $\mathrm{GD}(K)$ is the union $\sqcup_{j=1}^{m} S_j^1$ with chords connecting points s_1, s_2 such that $\mathrm{pr}_{xz}(f(s_1)) = \mathrm{pr}_{xz}(f(s_2))$. Gauss diagrams $\mathrm{GD}_1, \mathrm{GD}_2$ are equivalent if there is an orientation preserving diffeomorphism of $\sqcup_{j=1}^{m} S_j^1$ such that the endpoints of any chord of GD_1 map onto the endpoints of a chord of GD_2 and vice versa.* ∎

Definition 4.8 *For each of 2 types of oriented triple points, the co-orientation of $\Sigma_{\ast}$ is defined in terms of the Gauss diagrams of the corresponding links $K_{\pm}$ in Fig. 4.6. Assume that, while $t \in S_t^1$ increases, the point $\mathrm{rot}_t(K) \in \mathrm{SL}$ passes through $\Sigma_{\ast}$ from the negative side to the positive one. Then associate to the corresponding triple vertex of $\mathrm{TG}(K)$ the positive sign $+$, otherwise take the negative sign $-$.*
The co-orientations of $\Sigma_{\times}$, $\Sigma_{\curlyvee}$ and of $\Sigma_{\curlywedge}$ are also defined in Fig. 4.6. ∎

Look at the trefoil K in Fig. 4.2 and its trace graph $\mathrm{TG}(K)$ in Fig. 4.3. Consider the first triple vertex of $\mathrm{TG}(K)$ at the critical moment $t_1 \in (\pi/4, \pi/2)$. The knot $\mathrm{rot}_{\pi/4}(K)$ is on the positive side of $\Sigma_{\ast}$, while $\mathrm{rot}_{\pi/2}(K)$ is on the negative side of $\Sigma_{\ast}$, i.e. the first triple vertex has the positive sign. At the second triple vertex for $t_2 \in (\pi/2, 3\pi/4)$, the knot $\mathrm{rot}_t(K)$ goes from the negative side to the positive side. So the second triple point also the positive sign.

4.3 Generic links, equivalences, loops and homotopies

4.3.1 The canonical loop of a link and generic links

Generic links will be defined as the most generic points in the space SL of all links $K \subset V$ with respect to the rotation rot_t of the solid torus V.

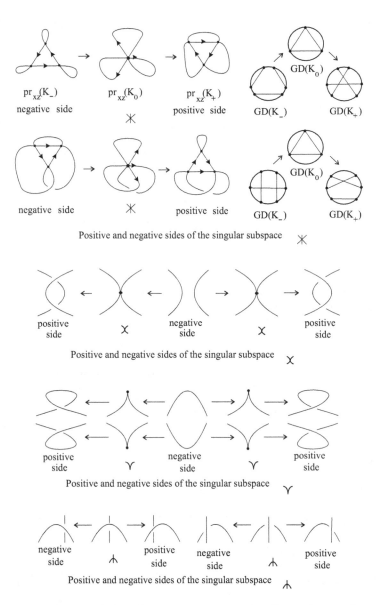

Figure 4.6: How to define the co-orientations of codimension 1 subspaces

Definition 4.9 *The* canonical *loop* $\mathrm{CL}(K) \subset \mathrm{SL}$ *of a smooth link* $K \subset V$ *is the union of the rotated links* $\mathrm{rot}_t(K) \in \mathrm{SL}$ *over all* $t \in S_t^1$.
A link $K \subset V$ *is* generic *if there are finitely many* $t_1, \ldots, t_k \in S_t^1$ *such that*

- *for all* $t \notin \{t_1, \ldots, t_k\}$, *the links* $\mathrm{rot}_t(K)$ *are general, i.e.* $\mathrm{rot}_t(K) \in \Sigma^{(0)}$;

- $\mathrm{CL}(K)$ *transversally intersects* $\Sigma_{\times} \cup \Sigma_{\chi} \cup \Sigma_{\curlyvee} \cup \Sigma_{\curlywedge}$ *at each* $t \in \{t_1, \ldots, t_k\}$.

Denote by $\Omega^{(0)} \subset \mathrm{SL}$ *the subspace of all generic links in* V. ∎

The canonical loop is the analogue of Gramain's loop in the case of knots in the solid torus.

Morse modifications of index 1 would change the trace graph dramatically. Luckily following Lemma 4.2 shows that they can not occur under strong equivalence. More exactly Lemma 4.2 shows that the canonical loop $\mathrm{CL}(K)$ never touches the subspace $\Sigma_{\chi} \cup \Sigma_{\curlyvee} \cup \Sigma_{\curlywedge}$ for any link K. Therefore the transversality from the last condition of Definition 4.9 is relevant only for the subspace $\Sigma_{\times}$.

Lemma 4.2 (Main topological lemma) *For any link* $K \subset V$, *the canonical loop* $\mathrm{CL}(K)$ *does not touch the subspace* $\Sigma_{\chi} \cup \Sigma_{\curlyvee} \cup \Sigma_{\curlywedge}$. *More formally, if* $K \in \Sigma_{\gamma}$ *for* $\gamma = \times, \curlyvee, \curlywedge$, *the links* $\mathrm{rot}_{\pm\varepsilon}(K)$ *are on different sides of* Σ_{γ} *for small* $\varepsilon > 0$.

Proof 4.1 *For the subspaces* Σ_{χ} *and* $\Sigma_{\curlyvee}$, *the projections of two small arcs with a tangent point (respectively, a cusp) are interchanged under the rotation.*

One easily sees in Fig. 4.6 that the links $\mathrm{rot}_{\pm\varepsilon}(K)$ *are on different sides of* Σ_{χ} *and* $\Sigma_{\curlyvee}$, *respectively, since the tangent at* p *is not horizontal, i.e. not orthogonal to the vertical axis* S_z^1. *The argument for* $\Sigma_{\curlywedge}$ *is the same, since one tangent at the critical crossing is not horizontal, see Fig. 4.6.*

Example 4.1 *The canonical loop* $\mathrm{CL}(K)$ *of a knot* $K \subset V$ *can touch the subspace* $\Sigma_{\times}$. *Consider the three arcs* $J_1, J_2, J_3 \subset \mathbb{R}^3$ *defined by (see Fig. 4.7 below)*

$$\begin{cases} x_1 = \tau u^2, \\ y_1 = 0, \\ z_1 = u; \end{cases} \qquad \begin{cases} x_2 = u, \\ y_2 = -1, \\ z_2 = u; \end{cases} \qquad \begin{cases} x_3 = -u, \\ y_3 = 1, \\ z_3 = u; \end{cases} \qquad u \in \mathbb{R}, \ \tau > 0.$$

Under the composition $\mathrm{pr}_{xz} \circ \mathrm{rot}_t$, *the arcs* J_1, J_2, J_3 *map to the following ones:*

$$x_1(t) = \tau z_1^2 \cos t, \quad x_2(t) = z_2 \cos t + \sin t, \quad x_3(t) = -z_3 \cos t - \sin t,$$

where z_1, z_2, z_3 *are constants. For small* $t = \varepsilon > 0$, *the double crossing* $p_{23} = \mathrm{pr}_{xz}(\mathrm{rot}_\varepsilon(J_2)) \cap \mathrm{pr}_{xz}(\mathrm{rot}_\varepsilon(J_3))$ *has the coordinates* $x = 0$, $z = -\tan \varepsilon$. *Then* p_{23} *is at the left of the first rotated arc* $x_1(t) = \tau z_1^2 \cos t$ *with respect to* X.

For $t = -\varepsilon < 0$, *the crossing with* $x = 0$, $z = \tan \varepsilon$ *is also at the left of the first arc. Take a knot* $K \in \Sigma_{X}$ *containing small parts of the arcs described above. Then* $\mathrm{rot}_{\pm\varepsilon}(K)$ *are on the same side of* Σ_{X}. *This means that, under the rotation of* K, *Reidemeister move III is not performed for the diagram* $\mathrm{pr}_{xz}(\mathrm{rot}_t(K))$.

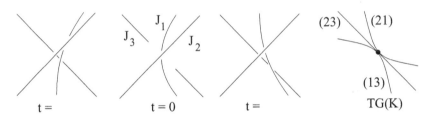

Figure 4.7: $\mathrm{CL}(K)$ may touch the subspace of triple points

4.3.2   Codimension 2 singularities and generic equivalences

Classical codimension 2 singularities of plane curves were described by David [15, List II on p. 561], namely the rumphoidal cusp ⌐ (the A_4 singularity in Arnold's notations), intersected cusp ⋎ (D_5), tangent triple point ⋇ (D_4), cubic tangency ⌐ (A_5) and ordinary quadruple point ✳ (X_9). We need to distinguish more refined singularities since the canonical loop of a link may not be transversal to some singular subspace, e.g. it is transversal to the codimension 2 subspace of horizontal cusps $\Sigma_{\_\prime}$, but not to the codimension 1 subspace of all cusps $\Sigma_{Y} \cup \Sigma_{\_\prime}$. All tangents below are not *horizontal* unless stated otherwise.

Definition 4.10 *(codimension 2 singularities of link diagrams)*
Let D be a diagram, i.e. the image of a smooth map $g : \sqcup_{j=1}^{m} S_j^1 \to A_{xz}$.

⁎ *A* quadruple point *of D is a transversal intersection p of 4 arcs.*

⤫ *A* tangent triple point *of D is an intersection p of 3 arcs, the first two arcs have a simple tangency and do not touch the third arc.*

⤬ *An* intersected cusp *of D is an intersection of 2 arcs, where the first arc has an ordinary cusp whose vector $(\ddot{x}, \ddot{z})$ does not touch the second arc.*

⨍ *A* cubic tangency *is an intersection of 2 arcs given locally by $u = 0$, $u = v^3$.*

⟍ *A* ramphoidal cusp *is the singular point of an arc, given locally by $u^2 = v^5$.*

≺ *A* horizontal cusp *is an ordinary cusp with horizontal tangent.*

⋊ *A* mixed tangency *is a simple tangency with a horizontal tangent such that one of the extrema is a maximum, another one is a minimum.*

⋀ *An* extreme tangency *is a simple tangency with a horizontal tangent such that* **both** *extrema are either maxima or minima.*

⋀,⋀ *A* horizontal triple point *is a triple intersection, where the tangent line of the first arc is horizontal, the tangent lines of the other arcs are not horizontal.*

Given a singularity $\delta \in \{ ⁎, ⤫, ⤬, ⨍, ⟍, ⋀, ⋀, ≺, ⋊, ⋀ \}$, denote by Σ_δ the union of all links $K \subset V$ such that the diagram $\mathrm{pr}_{xz}(K)$ is general outside δ. Set

$$\Sigma^{(2)} = \Sigma_⁎ \cup \Sigma_⤫ \cup \Sigma_⤬ \cup \Sigma_⨍ \cup \Sigma_⟍ \cup \Sigma_⋀ \cup \Sigma_⋀ \cup \Sigma_≺ \cup \Sigma_⋊ \cup \Sigma_⋀ \subset \mathrm{SL}. \quad ■$$

Lemma 4.3 *The singular subspace $\Sigma^{(2)}$ has codimension 2 in the space* SL.

Proof 4.2 *We use multi jets of maps* $(x(r), z(r)) : \mathbb{R} \to \mathbb{R}^2$ *defining a diagram D. Fixing 4 points r_1, r_2, r_3, r_4, each singularity δ from Definition 4.10 can be described in terms of 4-tuple 3-jets from the space $J^3_{[4]}(\mathbb{R}, \mathbb{R}^2)$, where each point has the 36 coordinates:*

$$
J^3_{[4]}(\mathbb{R}, \mathbb{R}^2) : \quad \left\{ r_i; \quad \begin{array}{ll} x(r_i), \quad z(r_i); & \dot{x}(r_i), \quad \dot{z}(r_i); \\ \ddot{x}(r_i), \quad \ddot{z}(r_i); & \dddot{x}(r_i), \quad \dddot{z}(r_i); \end{array} \quad i = 1, 2, 3, 4. \right.
$$

The jets over all $K \in \Omega_\delta$ form the finite dimensional subspace $\tilde{\Sigma}_\delta \subset J^3_{[4]}(\mathbb{R}, \mathbb{R}^2)$. *A simple tangency of 2 arcs at r_i, r_j is described by* $\Gamma_{ij} = \begin{vmatrix} \dot{x}(r_i) & \dot{x}(r_j) \\ \dot{z}(r_i) & \dot{z}(r_j) \end{vmatrix} = 0$. *The frequent inequality $\dot{z}(r_i) \neq 0$ below says that the tangent of D at r_i is not horizontal. The string $r_1 \neq r_2 \neq r_3 \neq r_4$ will mean that r_1, r_2, r_3, r_4 are pairwise disjoint.*

$\tilde{\Sigma}_{\divideontimes} \left\{ \begin{array}{l} x(r_1) = x(r_2) = x(r_3) = x(r_4), \quad r_1 \neq r_2 \neq r_3 \neq r_4, \\ z(r_1) = z(r_2) = z(r_3) = z(r_4), \quad \dot{z}(r_i) \neq 0, \; \Gamma_{ij} \neq 0, \; i \neq j; \end{array} \right.$

$\tilde{\Sigma}_{\mathcal{X}} \left\{ \begin{array}{l} x(r_1) = x(r_2) = x(r_3), \quad r_1 \neq r_2 \neq r_3 = r_4, \; \dot{z}(r_i) \neq 0, \\ z(r_1) = z(r_2) = z(r_3), \quad \Gamma_{12} = 0, \; \Gamma_{23} \neq 0, \; \Gamma_{13} \neq 0; \end{array} \right.$

$\tilde{\Sigma}_{\curlyvee} \left\{ \begin{array}{ll} x(r_1) = x(r_2), z(r_1) = z(r_2), & \dot{z}(r_1) \neq 0, \dot{z}(r_2) \neq 0, \; \begin{vmatrix} \ddot{x}(r_1) & \dot{x}(r_2) \\ \ddot{z}(r_1) & \dot{z}(r_2) \end{vmatrix} \neq 0; \\ \dot{x}(r_1) = \dot{z}(r_1) = 0, & r_1 \neq r_2 = r_3 = r_4, \end{array} \right.$

$\tilde{\Sigma}_{\curlyvee} \left\{ \begin{array}{l} x(r_1) = x(r_2), z(r_1) = z(r_2), \quad \begin{vmatrix} \ddot{x}(r_1) & \dot{x}(r_2) \\ \ddot{z}(r_1) & \dot{z}(r_2) \end{vmatrix} = 0, \; \begin{vmatrix} \dddot{x}(r_1) & \dot{x}(r_2) \\ \dddot{z}(r_1) & \dot{z}(r_2) \end{vmatrix} \neq 0; \\ r_1 \neq r_2 = r_3 = r_4, \Gamma_{12} = 0, \\ \dot{z}(r_1) \neq 0, \dot{z}(r_2) \neq 0, \end{array} \right.$

$\tilde{\Sigma}_{\gamma} \left\{ \begin{array}{ll} \dot{x}(r_1) = \dot{z}(r_1) = 0, \quad \dddot{z}(r_1) \neq 0 & \begin{vmatrix} \ddot{x}(r_1) & \dddot{x}(r_1) \\ \ddot{z}(r_1) & \dddot{z}(r_1) \end{vmatrix} = 0; \\ r_1 = r_2 = r_3 = r_4, \end{array} \right.$

$\tilde{\Sigma}_{\divideontimes} \cup \tilde{\Sigma}_{\divideontimes} \left\{ \begin{array}{l} x(r_1) = x(r_2) = x(r_3), \quad \dot{z}(r_1) = 0, \ddot{z}(r_1) \neq 0, \; r_1 \neq r_2 \neq r_3 = r_4, \\ z(r_1) = z(r_2) = z(r_3), \quad \dot{z}(r_2) \neq 0, \dot{z}(r_3) \neq 0, \; \Gamma_{ij} \neq 0, i \neq j; \end{array} \right.$

$\tilde{\Sigma}_{\prec} \quad \{ \quad \dot{x}(r_1) = \ddot{z}(r_1) = \ddot{z}(r_1) = 0, \quad r_1 = r_2 = r_3 = r_4, \quad \dddot{z}(r_1) \neq 0;$

$\tilde{\Sigma}_{\asymp} \cup \tilde{\Sigma}_{\wedge} \left\{ \begin{array}{l} z(r_1) = z(r_2), \quad \dot{z}(r_1) = \dot{z}(r_2) = 0, \quad \ddot{z}(r_1) \neq 0, \\ x(r_1) = x(r_2), \quad r_1 \neq r_2 = r_3 = r_4, \quad \ddot{z}(r_2) \neq 0; \end{array} \right.$

The conditions above can be obtained using classical normal forms of the singularities, e.g. the ramphoidal cusp $\rotatebox{90}{$\prec$}$ is a degeneration of the ordinary cusp $\curlyvee$ clearly given by $\dot{x}(r) = \dot{z}(r) = 0$. Locally one has $(x, z) = (a_2r^2 + a_3r^3 + \dots, b_2r^2 + b_3r^3 + \dots)$, $b_2 \neq 0$, which is (left) equivalent to $(x, z) = ((a_3 - b_3a_2/b_2)r^3 + \dots, b_2r^2 + \dots)$, hence $a_2b_3 = b_2a_3$, i.e. the vectors $(\ddot{x}(r), \ddot{z}(r))$ and $(\dddot{x}(r), \dddot{z}(r))$ are collinear.

Each subspace $\tilde{\Sigma}_\delta$ is defined by 6 equations, hence $\text{codim}\,\tilde{\Omega}_\delta = 6$ in $J_{[4]}^3(\mathbb{R}, \mathbb{R}^3)$. The subspaces Σ_δ from Definition 4.10 map to the corresponding subspaces $\tilde{\Sigma}_\delta \subset J_{[4]}^3(\mathbb{R}, \mathbb{R}^2)$ by adding 4 points r_1, r_2, r_3, r_4 on a diagram. When we forget these points, the codimension decreases by 4, i.e. $\text{codim}\,\Sigma_\delta = 2$ in the space SL.

Definition 4.11 *Let $\Omega_{\overline{\times}}$ be the set of all links failing to be generic due to exactly one tangency of* $\overset{\curvearrowright}{\text{CL}}(K)$ *with the codimension 1 subspace $\Sigma_{\times}$.*

Given $\delta \in \{\text{\ding{89}}, \mathcal{X}, \mathcal{Y}, \mathcal{J}, \rotatebox{90}{$\prec$}, \text{\ding{73}}, \mathcal{A}, \prec, \times, \mathcal{M}\}$, let Ω_δ consist of all links K failing to be generic because of exactly one transversal intersection of $\text{CL}(K)$ with Σ_δ. Set

$$\Omega^{(1)} = \Omega_{\text{\ding{89}}} \cup \Omega_{\mathcal{X}} \cup \Omega_{\mathcal{Y}} \cup \Omega_{\mathcal{J}} \cup \Omega_{\rotatebox{90}{$\prec$}} \cup \Omega_{\text{\ding{73}}} \cup \Omega_{\mathcal{A}} \cup \Omega_{\prec} \cup \Omega_{\times} \cup \Omega_{\mathcal{M}} \cup \Omega_{\overline{\times}}.$$

A generic equivalence is a smooth path $F : [0, 1] \to \text{SL}$ intersecting transversally the subspace $\Omega^{(1)}$, i.e. there are finitely many $r_1, \dots, r_k \in [0, 1]$ such that

- *the links $F(r) \in \text{SL}$ are generic for all $r \notin \{r_1, \dots, r_k\}$;*

- *the canonical loop $\text{CL}(F(r))$ transversally intersects $\Omega^{(1)}$ for $r = r_1, \dots, r_k$.* $\blacksquare$

Lemma 4.4 **(a)** *The subspace $\Omega^{(1)}$ has codimension 1 in the space SL.*

(b) *The subspace $\Omega^{(0)}$ is open and dense in the space SL.*

Proof 4.3 **(a)** *Choose a link $K \subset V$ given by an embedding $f : \bigsqcup_{j=1}^{m} S_j^1 \to V$ that fails to be generic due to exactly one singularity δ from Definition 4.11. These singularities were introduced using the rotation of the solid torus V. So we describe them in terms of maps $\mathbb{R} \to \mathbb{R}^3$, not $\mathbb{R} \to \mathbb{R}^2$ as in the proof of Lemma 4.3.*

There is a 4-tuple $(r_1, r_2, r_3, r_4) \in (\sqcup_{j=1}^m S_j^1)^4$ defining the chosen singularity of $K \in \Omega_\delta$. The 4-tuple 3-jet $j_{[4]}^3 f(r_1, r_2, r_3, r_4)$ is a point in $J_{[4]}^3(\mathbb{R}, \mathbb{R}^3)$. These points over all $K \in \Omega_\delta$ form the finite dimensional subspace $\tilde\Omega_\delta \subset J_{[4]}^3(\mathbb{R}, \mathbb{R}^3)$.

We check that $\tilde\Omega_\delta$ has codimension 5 in $J_{[4]}^3(\mathbb{R}, \mathbb{R}^3)$. Denote by $x(r)$, $y(r)$, $z(r)$ the compositions of $f : \sqcup_{j=1}^m S_j^1 \to V \subset \mathbb{R}^3$ and the projections to the coordinate axes. Then the 4-tuple 3-jet of K is determined by the following 52 quantities.

$$J_{[4]}^3(\mathbb{R}, \mathbb{R}^3) : \left\{ r_i; \begin{array}{l} x(r_i), \ y(r_i), \ z(r_i); \quad \dot x(r_i), \ \dot y(r_i), \ \dot z(r_i); \\ \ddot x(r_i), \ \ddot y(r_i), \ \ddot z(r_i); \quad \dddot x(r_i), \ \dddot y(r_i), \ \dddot z(r_i); \end{array} \quad i = 1, 2, 3, 4. \right.$$

For $i, j \in \{1, 2, 3, 4\}$, $i \neq j$, introduce the differences

$$\Delta x_{ij} = x(r_i) - x(r_j), \quad \Delta y_{ij} = y(r_i) - y(r_j), \quad \Delta z_{ij} = z(r_i) - z(r_j).$$

Points $f(r_i), f(r_j), f(r_k) \in K$ project to the same point under pr_{xz} : $\mathrm{rot}_t(K) \to A_{xz} \times \{t\}$ for some t if and only if $z(r_i) = z(r_j) = z(r_k)$, $\begin{vmatrix} \Delta x_{ij} & \Delta x_{jk} \\ \Delta y_{ij} & \Delta y_{jk} \end{vmatrix} = 0$. The last determinant is (up to the sign) the area of the triangle with the vertices $(x(r_i), y(r_i))$, $(x(r_j), y(r_j))$, $(x(r_k), y(r_k))$ in the horizontal plane $\{z(r_i) = z(r_j) = z(r_k)\}$.

Set $\Delta_{ij} = \begin{vmatrix} \dot x(r_i) & \dot x(r_j) & \Delta x_{ij} \\ \dot y(r_i) & \dot y(r_j) & \Delta y_{ij} \\ \dot z(r_i) & \dot z(r_j) & \Delta z_{ij} \end{vmatrix}$. The diagram $\mathrm{pr}_{xz}(\mathrm{rot}_t(K))$ contains two arcs having a simple tangency at $r = r_i$, $r = r_j$ and some t if and only if $z(r_i) = z(r_j)$ and $\Delta_{ij} = 0$, i.e. the straight line through $f(r_i), f(r_j) \in K$ lies in the plane spanned by the tangent vectors of K at $r = r_i$ and $r = r_j$.

We describe analytically the subspaces $\tilde\Omega_\delta$ associated to the singularities

$$\ast, \times, \curlyvee, \curlywedge, \gtrdot, \ast, \nearrow, \ltimes, \prec, \curlywedge, \overline{\ast}.$$

$$\tilde\Omega_{\ast} \begin{cases} z(r_1) = z(r_2) = z(r_3) = z(r_4), & \begin{vmatrix} \Delta x_{12} & \Delta x_{23} \\ \Delta y_{12} & \Delta y_{23} \end{vmatrix} = \begin{vmatrix} \Delta x_{12} & \Delta x_{24} \\ \Delta y_{12} & \Delta y_{24} \end{vmatrix} = 0, \\ r_1 \neq r_2 \neq r_3 \neq r_4, & \\ \dot z(r_i) \neq 0, \ i = 1, 2, 3, 4, & \Delta_{ij} \neq 0, \ i, j \in \{1, 2, 3, 4\}, \ i \neq j; \end{cases}$$

$$\tilde\Omega_{\times} \begin{cases} z(r_1) = z(r_2) = z(r_3), & \begin{vmatrix} \Delta x_{12} & \Delta x_{23} \\ \Delta y_{12} & \Delta y_{23} \end{vmatrix} = 0, \\ r_1 \neq r_2 \neq r_3 = r_4, & \\ \dot z(r_i) \neq 0, \ i = 1, 2, 3, & \Delta_{12} = 0, \ \Delta_{23} \neq 0, \ \Delta_{13} \neq 0; \end{cases}$$

$$\tilde{\Omega}_{\curlyvee}\begin{cases} z(r_1) = z(r_2), \\ \dot{z}(r_1) = 0,\ \ddot{z}(r_1) \neq 0, \\ \dot{z}(r_2) \neq 0, \end{cases} \begin{vmatrix} \dot{x}(r_1) & \dot{x}(r_2) \\ \dot{y}(r_1) & \dot{y}(r_2) \end{vmatrix} = 0, \quad \begin{vmatrix} \ddot{x}(r_1) & \dot{x}(r_2) & \Delta x_{12} \\ \ddot{y}(r_1) & \dot{y}(r_2) & \Delta y_{12} \\ \ddot{z}(r_1) & \dot{z}(r_2) & \Delta z_{12} \end{vmatrix} \neq 0;$$
$$r_1 \neq r_2 = r_3 = r_4,$$

$$\tilde{\Omega}_{\curlywedge}\begin{cases} z(r_1) = z(r_2),\quad \Delta_{12} = 0, \\ r_1 \neq r_2 = r_3 = r_4, \\ \dot{z}(r_1) \neq 0,\quad \dot{z}(r_2) \neq 0, \end{cases} \begin{vmatrix} \ddot{x}(r_1) & \ddot{x}(r_2) & \Delta x_{12} \\ \ddot{y}(r_1) & \ddot{y}(r_2) & \Delta y_{12} \\ \ddot{z}(r_1) & \ddot{z}(r_2) & \Delta z_{12} \end{vmatrix} = 0;$$

$$\tilde{\Omega}_{\curvearrowleft}\begin{cases} \dot{z}(r_1) = 0,\quad \ddot{z}(r_1) \neq 0, \\ r_1 = r_2 = r_3 = r_4, \end{cases} \frac{\dddot{x}(r_1)}{\ddot{x}(r_1)} = \frac{\dddot{y}(r_1)}{\ddot{y}(r_1)} = \frac{\dddot{z}(r_1)}{\ddot{z}(r_1)}.$$

The last equations with 3 fractions mean that the vectors of the 2nd and 3rd derivatives are collinear, which corresponds to the similar condition for $\tilde{\Sigma}_{\curvearrowleft}$ in the proof of Lemma 4.3. If a denominator is zero, the numerator must be also zero.

$$\tilde{\Omega}_{\mathbb{X}} \cup \tilde{\Omega}_{\cancel{X}}\begin{cases} z(r_1) = z(r_2) = z(r_3),\quad r_1 \neq r_2 \neq r_3 = r_4, \\ \dot{z}(r_1) = 0,\ \ddot{z}(r_1) \neq 0,\ \dot{z}(r_2) \neq 0,\ \dot{z}(r_3) \neq 0, \end{cases} \begin{vmatrix} \Delta x_{12} & \Delta x_{23} \\ \Delta y_{12} & \Delta y_{23} \end{vmatrix} = 0;$$

$$\tilde{\Omega}_{\curlyeqprec}\ \{\ \dot{z}(r_1) = \ddot{z}(r_1) = 0,\quad r_1 = r_2 = r_3 = r_4,\quad \dddot{z}(r_1) \neq 0;$$

$$\tilde{\Omega}_{\curlyeqprec} \cup \tilde{\Omega}_{\curlywedge}\ \begin{cases} z(r_1) = z(r_2),\quad\ r_1 \neq r_2 = r_3 = r_4, \\ \dot{z}(r_1) = \dot{z}(r_2) = 0,\ \ddot{z}(r_1) \neq 0,\ \ddot{z}(r_2) \neq 0. \end{cases}$$

If $\dot{z}(r_i) \neq 0$, then locally r_i can be considered as a function of z, hence any function of (several) r_i can be differentiated with respect to z. Below the tangency with $\Sigma_{\mathbb{X}}$ means that the derivative of the vanishing determinant $\Delta = \begin{vmatrix} \Delta x_{12} & \Delta x_{23} \\ \Delta y_{12} & \Delta y_{23} \end{vmatrix}$ defining a triple point under the projection $\mathrm{pr}_{xz} : \mathrm{rot}_t(K) \to \mathbb{A}_{xz} \times \{t\}$ also vanishes.

$$\tilde{\Omega}_{\overline{\mathbb{X}}}\begin{cases} z(r_1) = z(r_2) = z(r_3),\quad r_1 \neq r_2 \neq r_3 = r_4 \quad \Delta_{ij} \neq 0, \\ \Delta = \dfrac{d}{dz}\Delta = 0,\ \dfrac{d^2}{dz^2}\Delta \neq 0,\ \Delta = \begin{vmatrix} \Delta x_{12} & \Delta x_{23} \\ \Delta y_{12} & \Delta y_{23} \end{vmatrix}\ \begin{matrix} i \neq j, \\ \dot{z}(r_i) \neq 0. \end{matrix} \end{cases}$$

Generic inequalities $\dfrac{dg}{dz} \neq 0$ should be added to the descriptions above for each condition $g = 0$, which guarantees no tangency of the canonical loop with the corresponding subspace Σ_δ. In important cases like $\tilde{\Omega}_{\curvearrowleft}$ we explicitly accompanied $\dot{z}(r_1) = 0$ with $\ddot{z}(r_2) \neq 0$ equivalent to $\dfrac{\dot{z}(r_1(z))}{dz} \neq 0$, but also every equation like $z(r_1) = z(r_2)$ should be accompanied with $\dfrac{dz(r_1(z))}{dz} \neq \dfrac{dz(r_2(z))}{dz}$.

Each subspace $\tilde{\Omega}_\delta$ is defined by 5 equations, hence codim $\tilde{\Omega}_\delta = 5$ in $J^3_{[4]}(\mathbb{R}, \mathbb{R}^3)$. The subspaces Ω_δ introduced geometrically in Definition 4.11 correspond to $\tilde{\Omega}_\delta$ by adding 4 points r_1, r_2, r_3, r_4 on a link. When we forget about these points the codimension decreases by 4, i.e. codim $\Omega_\delta = 1$ in the space SL of all links $K \subset V$.

(b) The conditions of Definition 4.11 define an open subset of SL whose complement is clearly the closure of the codimension 1 subspace $\Omega^{(1)}$.

The following result similar to Proposition 4.1 follows from Lemma 4.4 since by Theorem 4.3 any isotopy in the space SL of links can be approximated by a path transversally intersecting the singular subspace $\Omega^{(1)} \subset$ SL.

Proposition 4.2 (a) *Any smooth link can be approximated by a generic link.*

(b) *Any smooth equivalence of links can be approximated by a generic one.*

4.3.3   Generic loops and generic homotopies in the space of links

A *loop* of links $\{K_t\} \subset$ SL means a *smooth* loop, i.e. a smooth map $S^1_t \to$ SL. Generic loops provide a suitable generalization of the canonical loop.

Definition 4.12 *A smooth loop of links* $\{K_t\} \subset$ SL, $t \in S^1_t$, *is called* generic *if there are finitely many critical moments* $t_1, \ldots, t_k \in S^1_t$ *such that*

- *the link* K_t *maps to* $K_{t+\pi}$ *under the rotation through* π *for every* $t \in S^1_t$;

- *for all* $t \notin \{t_1, \ldots, t_k\}$, *the links* K_t *are general, i.e.* $K_t \in \Sigma^{(0)}$;

- $\{K_t\}$ *transversally intersects* $\Sigma_{\times} \cup \Sigma_{\rightthreetimes} \cup \Sigma_{\curlyvee} \cup \Sigma_{\curlywedge}$ *at each* $t = t_1, \ldots, t_k$. ∎

Due to Lemmas 4.1, 4.4 any loop can be approximated by a generic loop. But a generic loop may be too trivial. For instance, a loop $S^1_t \to$ SL contractible to a generic link through generic links carries information about only one diagram. More interesting objects are generic loops homotopic to canonical loops.

Definition 4.13 *A smooth family* $\{L_s\}$ *of loops,* $s \in [0, 1]$, *is called a* generic homotopy *if there are finitely many critical moments* $s_1, \ldots, s_k \in [0, 1]$ *such that*

- *for* $s \notin \{s_1, \ldots, s_k\}$, *the loop* L_s *is generic in the sense of Definition 4.12;*

- *for each* $s \in \{s_1, \ldots, s_k\}$, *the loop* L_s *fails to be generic since either* L_s *transversally intersects* $\Sigma^{(2)}$ *or* L_s *touches* $\Sigma_{\bowtie}$ *at exactly one point.*

∎

Lemma 4.5 **(a)** *The canonical loop of any generic link is a generic loop.*

(b) *Any generic equivalence* $\{K_s\}$, $s \in [0, 1]$, *of links provides the generic homotopy of loops* $\{\mathrm{CL}(K_s)\}$ *of links.*

(c) *If canonical loops* $\mathrm{CL}(K_0)$ *and* $\mathrm{CL}(K_1)$ *of generic links* K_0 *and* K_1 *are generically homotopic then* K_0 *and* K_1 *are generically equivalent.*

Proof 4.4 **(a)** *The canonical loop of any link is symmetric in the sense that* $\mathrm{rot}_t(K)$ *maps to* $\mathrm{rot}_{t+\pi}(K)$ *under the rotation through* π *for every* $t \in S^1_t$. *The other conditions of Definition 4.12 correspond to the conditions of Definition 4.9.*

(b) *Compare Definition 4.11 with Definitions 4.12 and 4.13.*

(c) *Let* $\{L_s\}$, $s \in [0, 1]$, *be a generic homotopy between* $\mathrm{CL}(K_0)$ *and* $\mathrm{CL}(K_1)$. *The loops* L_s *can be represented by a cylinder* $S^1_t \times [0, 1]$ *mapped to the space* SL. *Take a smooth path connecting* K_0 *and* K_1 *inside the cylinder. This smooth equivalence can be approximated by a generic one due to Proposition 4.2b.*

By Lemma 4.5 the classification of links reduces to their canonical loops.

Proposition 4.3 *Generic links are generically equivalent in* V *if and only if their canonical loops are generically homotopic in the space* SL *of all links* $K \subset V$.

4.4   Through codimension 2 singularities

4.4.1   Versal deformations of codimension 2 singularities

To understand what happens when the canonical loop of a link passes through the singular subspace $\Sigma^{(2)}$, we study bifurcation diagrams of codimension 2 singularities.

Lemma 4.6 *The codimension 2 singularities from Definition 4.10 have the normal forms in the table below, where r is the parameter on the curve and*

- $\mathcal{A}_e$ *is the extended right-left equivalence, i.e. diffeomorphisms of $\mathbb{R}^2$ don't fix 0;*

- $\mathcal{A}_z$ *is the restricted right-left equivalence such that left diffeomorphisms of $\mathbb{R}^2$ have the form $(g(x,z), h(z))$, where $g(x,z) : \mathbb{R}^2 \to \mathbb{R}$, $h(z) : \mathbb{R} \to \mathbb{R}$ are diffeomorphisms.*

$\divideontimes$, $\mathcal{A}_e$	$\{x = 0,\ z = r\}, \{x = r,\ z = r\}, \{x = -r,\ z = r\}, \{x = er,\ z = r\}$	
$\mathcal{X}$, $\mathcal{A}_e$	$\{x = r^2,\ z = r\}$, $\{x = 0,\ z = r\}$, $\{x = r,\ z = r\}$	
$\curlyvee$, $\mathcal{A}_e$	$\{x = r^3,\ z = r^2\}$, $\{x = r,\ z = r\}$	
$\curlywedge$, $\mathcal{A}_e$	$\{x = r^3,\ z = r\}$, $\{x = 0,\ z = r\}$	
$\rightharpoondown$, $\mathcal{A}_e$	$\{x = r^5,\ z = r^2\}$	
$\prec$, $\mathcal{A}_z$	$\{x = r^2,\ z = r^3\}$	
$\bowtie$, $\mathcal{A}_z$	$\{x = r,\ z = r^2\}$, $\{x = r,\ z = -r^2\}$	
$\wedge\!\wedge$, $\mathcal{A}_z$	$\{x = r,\ z = -2r^2\}$, $\{x = r,\ z = -r^2\}$	
$\divideontimes$, $\mathcal{A}_z$	$\{x = r,\ z = -r^2\}$, $\{x = r,\ z = r\}$, $\{x = -r,\ z = r\}$	
$\not\!\divideontimes$, $\mathcal{A}_z$	$\{x = r,\ z = -r^2\}$, $\{x = r,\ z = r\}$, $\{x = 2r,\ z = r\}$	

Sketch 4.2 *The normal forms up to $\mathcal{A}_e$-equivalence are classical, e.g. the parameter $e \neq 0$ in the normal form of $\divideontimes$ (X_9) can not be skipped as the cross-ratio of 4 slopes is invariant under diffeomorphisms, see [57, Lemma 6.5]. The singularities $\bowtie$, $\prec$, $\wedge\!\wedge$, $\divideontimes$, $\not\!\divideontimes$ should be considered up to $\mathcal{A}_z$-equivalence respecting $\{z = \text{const}\}$, otherwise they don't have codimension 2, e.g. the normal form (r^2, r^3) of $\prec$ is not $\mathcal{A}_z$-equivalent to the normal form (r^3, r^2) of $\curlyvee$. Deducing new normal forms is similar, e.g. the horizontal cusp $\prec$ is defined by the conditions $\dot{x}(0) = \dot{z}(0) = \ddot{z}(0) = 0$, hence $x(r) = ar^2 + \ldots$, $z(r) = br^3 + \ldots$, which normalises to (r^2, r^3) as required.*

Mancini and Ruas [38] have shown that the group $\mathcal{A}_z$ from Lemma 4.6 is geometric in the sense of Damon [14]. So the standard technique of singularity theory can be applied to find versal deformations of corresponding codimension 1 singularities.

We consider horizontal triple points ⋌ and ⋌ separately, because the associated moves on trace graphs look slightly different in Figs. 4.11ix, 4.11x. A deformation of a germ $(x(r), z(r)) : \mathbb{R} \to \mathbb{R}^2$ with parameters a, b is a germ $F : \mathbb{R} \times \mathbb{R}^2 \to \mathbb{R}^2$ such that $F(r; 0, 0) \equiv (x(r), z(r))$. A deformation F is *versal* if any other deformation can be obtained from F by actions of the corresponding group $\mathcal{A}_e$ or $\mathcal{A}_z$.

The versality can be checked using the following tangent spaces at germs in the space of deformations. Let T^r be the *right* tangent space at a germ $(x(r), z(r))$ generated by the right diffeomorphisms $\mathbb{R} \to \mathbb{R}$, e.g. the right space T^r at (r^5, r^2) of ⋋ consists of $(5r^4 f(r), 2rf(r))$, where $f : \mathbb{R} \to \mathbb{R}$. Denote by T^l the *left* tangent space at a germ $(x(r), z(r))$ generated by the restricted left diffeomorphisms $(g(x, z), h(z)) : \mathbb{R}^2 \to \mathbb{R}^2$, where $g : \mathbb{R}^2 \to \mathbb{R}$, $h : \mathbb{R} \to \mathbb{R}$ are any germs. For instance, the left space T^l at (r^2, r^3) of ⋌ is formed by $(g(r^2, r^3), h(r^3)) = (a_1 + a_2 r^2 + a_3 r^3 + \ldots, b_1 + b_2 r^3 + \ldots)$. The *parameter* normal space N^p of a deformation $F(r; a, b)$ consists of linear combinations $c\dfrac{\partial F}{\partial a} + d\dfrac{\partial F}{\partial b}$ at $a = b = 0$, where c, d are constants, e.g. the space N^p of $(r^5 + ar^3 + br, r^2)$ consists of vectors $(cr^3 + dr, 0)$.

In the case of a multi-germ the right space T_i^r is associated to the independent right diffeomorphisms $f_i(r)$ around each point r_i. The left space T_i^l is generated by the same left diffeomorphisms at every r_i. The parameter space N_i^p is spanned by the derivatives along the parameters of the deformation at each r_i.

The following standard statement is a simple application of [1, Section I.8.2].

Proposition 4.4 *A deformation $F(r; a, b)$ of a multi-germ $(x(r), z(r)) : \mathbb{R} \to \mathbb{R}^2$ is versal if at every point r_i any germ $\mathbb{R} \to \mathbb{R}^2$ can be represented as a sum of vectors from the spaces T_i^r, T_i^l and N_i^p.*

Lemma 4.7 *The codimension 2 singularities from Definition 4.10 have the versal deformations with parameters a, b in the table below.*

$\divideontimes$, $\mathcal{A}_e$	$\{x=0, z=r\}, \{x=r+a, z=r\}, \{x=-r-b, z=r\}, \{x=er, z=r\}$
$\asymp$, $\mathcal{A}_e$	$\{x = r^2 - 2a, z = r\}, \{x = 0, z = r\}, \{x = r - b, z = r\}$
$\curlyvee$, $\mathcal{A}_e$	$\{x = r^3 - br, z = r^2\}, \{x = r - a, z = r\}$
$\curlyvee$, $\mathcal{A}_e$	$\{x = r^3 - 3br + a, z = r\}, \{x = 0, z = r\}$
$\gtrdot$, $\mathcal{A}_e$	$\{x = r^5 + ar^3 + br, z = r^2\}$
$\prec$, $\mathcal{A}_z$	$\{x = r^2, z = r^3 + ar^2 - br\}$
$\ltimes$, $\mathcal{A}_z$	$\{x = r, z = r^2 - b\}, \{x = r + a, z = -r^2\}$
$\curlywedge$, $\mathcal{A}_z$	$\{x = r, z = -2r^2 - b\}, \{x = r + a, z = -r^2\}$
$\divideontimes$, $\mathcal{A}_z$	$\{x = r, z = -r^2\}, \{x = r + a, z = r\}, \{x = -r - b, z = r\}$
$\divideontimes$, $\mathcal{A}_z$	$\{x = r, z = -r^2\}, \{x = r + a, z = r\}, \{x = r/2 - b, z = r\}$

Sketch 4.3 *Versal deformations of classical codimension 2 singularities $\gtrdot$ (A_4), $\curlyvee$ (D_5), $\asymp$ (D_4), $\curlyvee$ (A_5) and $\divideontimes$ (X_9) up to $\mathcal{A}_e$-equivalence were recently described by Wall [57, subsection 6.1]. The remaining cases follow from the table below.*

singularity	T_i^r	T_i^l	N_i^p
$\prec$	$(2rf(r), 3r^2 f(r))$	$(g(r^2, r^3), h(r^3))$	$(0, cr^2 - dr)$
$\ltimes$	$(f_1(r), 2rf_1(r))$	$(g(r, r^2), h(r^2))$	$(0, -d)$
	$(f_2(r), -2rf_2(r))$	$(g(r, -r^2), h(-r^2))$	$(c, 0)$
$\curlywedge$	$(f_1(r), -4rf_1(r))$	$(g(r, -2r^2), h(-2r^2))$	$(0, -d)$
	$(f_2(r), -2rf_2(r))$	$(g(r, -r^2), h(-r^2))$	$(c, 0)$
$\divideontimes$	$(f_1(r), -2rf_1(r))$	$(g(r, -r^2), h(-r^2))$	$(0, 0)$
	$(f_2(r), f_2(r))$	$(g(r, r), h(r))$	$(c, 0)$
	$(-f_3(r), f_3(r))$	$(g(-r, r), h(r))$	$(-d, 0)$
$\divideontimes$	$(f_1(r), -2rf_1(r))$	$(g(r, -r^2), h(-r^2))$	$(0, 0)$
	$(f_2(r), f_2(r))$	$(g(r, r), h(r))$	$(c, 0)$
	$(-f_3(r)/2, f_3(r))$	$(g(-r/2, r), h(r))$	$(-d, 0)$

Case vi *of a horizontal cusp* $\prec$. *By Proposition 4.4 we should prove that any germ* $(x(r), z(r)) : \mathbb{R} \to \mathbb{R}^2$ *can be represented as a sum of vectors from the spaces* T_1^r, T_1^l *and* N_1^p, *i.e. we solve the functional equations from the table* $x(r) = 2rf(r) + g(r^2, r^3)$ *and* $z(r) = -dr + cr^2 + 3r^2 f(r) + h(r^3)$, *which have one the of the possible solutions*

$$
\begin{cases}
d = -\dot{z}(0), \quad h(r^3) = z(0), \quad f(r) = \dfrac{z(r) - \dot{z}(0)r - z(0)}{3r^2} + \dfrac{\dot{x}(0)}{2} - \dfrac{\ddot{z}(0)}{6}, \\
c = \dfrac{\ddot{z}(0) - 3\dot{x}(0)}{2}, g(r^2, r^3) = x(r) - \dot{x}(0)r - 2\dfrac{z(r) - \ddot{z}(0)r^2/2 - \dot{z}(0)r - z(0)}{3r}.
\end{cases}
$$

Here h has only the constant term and $g(r^2, r^3)$ has no linear term in r, all other powers have the form $2j + 3k$ for some integers $j, k \geq 0$, e.g.

for a germ $(a_0 + a_1 r + a_2 r^2 + \ldots, \quad b_0 + b_1 r + b_2 r^2 + \ldots)$ one has

$f = a_1/2 + \ldots, g(x, z) = a_0 + a_2 x + \ldots, h = b_0, d = -b_1, c = (2b_2 - 3a_1)/2.$

Case vii *of a mixed tangency $\bowtie$. We prove that at each point r_i, $i = 1, 2$, any germ $(x_i, z_i) : \mathbb{R} \to \mathbb{R}^2$ can be represented as a sum of vectors from T_i^r, T_i^l, N_i^p, i.e. in terms of suitable c, d and f, g, h. Write down the equations from the table above.*

$$(\bowtie) \quad \begin{cases} x_1(r) = f_1(r) + g(r, r^2), & z_1(r) = 2rf_1(r) + h(r^2) - d, \\ x_2(r) = c + f_2(r) + g(r, -r^2), & z_2(r) = -2rf_2(r) + h(-r^2). \end{cases}$$

For a function $f(r)$ denote its constant term simply by f. The equations $z_1(r) = 2rf_1(r) + h(r^2) - d$ and $z_2(r) = 2rf_2(r) + h(-r^2)$ in degree 1 determine the constant terms f_1, f_2 of $f_1(r), f_2(r)$. Then system $(\bowtie)$ in degree 0 has a unique solution:

$$(\bowtie_0) \quad \begin{cases} x_1 = f_1 + g, & z_1 = h - d, \\ x_2 = c + f_2 + g, & z_2 = h. \end{cases} \quad \begin{cases} g = x_1 - f_1, & h = z_2, \\ c = x_2 - x_1 + f_1 - f_2, & d = z_2 - z_1. \end{cases}$$

For a function $f(r)$ define its odd and even part as $\text{Odd } f(r) = \dfrac{f(r) - f(-r)}{2}$,

$\text{Even } f(r) = \dfrac{f(r) + f(-r)}{2}$. *We look for solutions $g(x, z) = g_1(x) + g_2(z)$ and $h(z)$ such that $g_2(-z) = -g_2(z)$, $h(z) = h(-z)$. Split each equation of $(\bowtie)$:*

$$\begin{cases} \text{Odd } x_1(r) = \text{Odd } f_1(r) + \text{Odd } g_1(r), \text{ Even } z_1(r) = 2r\text{Odd } f_1(r) + h(r^2) - d, \\ \text{Even } x_1(r) = \text{Even } f_1(r) + \text{Even } g_1(r) + g_2(r^2), \text{Odd } z_1(r) = 2r\text{Even } f_1(r), \\ \text{Odd } x_2(r) = \text{Odd } f_2(r) + \text{Odd } g_1(r), \text{ Even } z_2(r) = -2r\text{Odd } f_2(r) + h(r^2) \\ \text{Even } x_2(r) = c + \text{Even } f_2(r) + \text{Even } g_1(r) - g_2(r^2), \text{Odd } z_2(r) = -2r\text{Even } f_2(r). \end{cases}$$

The resulting system has a solution below, where $\text{Even } z_1(r) - \text{Even } z_2(r) + d$ is divisible by r due to $(\bowtie_0)$. So the deformation is versal by Proposition 4.4.

$$\begin{cases} \text{Even } f_1(r) = \text{Odd } z_1(r)/2r, \quad \text{Even } f_2(r) = -\text{Odd } z_2(r)/2r, \\ \text{Odd } f_1(r) = (\text{Even } z_1(r) - \text{Even } z_2(r) + d)/4r + (\text{Odd } x_1(r) - \text{Odd } x_2(r))/2, \\ \text{Odd } f_2(r) = (\text{Even } z_1(r) - \text{Even } z_2(r) + d)/4r + (\text{Odd } x_2(r) - \text{Odd } x_1(r))/2, \\ \text{Even } g_1(r) = (\text{Even } x_1(r) + \text{Even } x_2(r) - \text{Odd } z_1(r)/2r + \text{Odd } z_2(r)/2r - c)/2, \\ \text{Odd } g_1(r) = (+\text{Even } z_2(r) - \text{Even } z_1(r) - d)/4r + (\text{Odd } x_1(r) + \text{Odd } x_2(r))/2, \\ g_2(r^2) = (\text{Even } x_1(r) - \text{Even } x_2(r) - \text{Odd } z_1(r)/2r - \text{Odd } z_2(r)/2r + c)/2, \\ h(r^2) = (\text{Even } z_1(r) + \text{Even } z_2(r) + d)/2 + r(\text{Odd } x_2(r) - \text{Odd } x_1(r)). \end{cases}$$

Case viii *of an extreme tangency* ⚹ *is similar to Case vii.*

Case ix *of a horizontal triple point* ⚹*. The table above gives*

$$(⚹) \quad \begin{cases} x_1(r) = f_1(r) + g(r, -r^2), & z_1(r) = -2rf_1(r) + h(-r^2), \\ x_2(r) = c + f_2(r) + g(r, r), & z_2(r) = f_2(r) + h(r), \\ x_3(r) = -d - f_3(r) + g(-r, r), & z_3(r) = f_3(r) + h(r). \end{cases}$$

The equation $z_1(r) = -2rf_1(r) + h(-r^2)$ *in degree 1 determines the constant term* f_1 *of the function* $f_1(r)$. *Then system* $(⚹)$ *in degree 0 has a unique solution.*

$$\begin{cases} x_1 = f_1 + g, & z_1 = h, \\ x_2 = c + f_2 + g, & z_2 = f_2 + h, \\ x_3 = -d - f_3 + g, & z_3 = f_3 + h. \end{cases}$$

$$\begin{cases} g = x_1 - f_1, & h = z_1, \\ f_2 = z_2 - z_1, & c = x_2 - x_1 + f_1 + z_1 - z_2, \\ f_3 = z_3 - z_1, & d = x_1 - f_1 - x_3 + z_1 - z_3. \end{cases}$$

We look for $g(x, z) = g_1(x) + g_2(z)$. *Apply elementary operations to* $(⚹)$

$$(⚹_1) \quad \begin{cases} 2rx_1(r) + z_1(r) = 2rg_1(r) + 2rg_2(-r^2) + h(-r^2), \\ x_2(r) - z_2(r) = c + g_1(r) + g_2(r) - h(r), \\ x_3(r) + z_3(r) = -d + g_1(-r) + g_2(r) + h(r). \end{cases}$$

The functions f_1, f_2, f_3 *can be expressed in terms of the solutions of* $(⚹_1)$. *Split the 1st equation of* $(⚹_1)$ *into the odd and even parts, then apply operations to* $(⚹_1)$:

$$2r\text{Odd}\, x_1(r) + \text{Even}\, z_1(r) = 2r\text{Odd}\, g_1(r) + h(-r^2),$$

$$(⚹_2) \qquad 2r\text{Even}\, x_1(r) + \text{Odd}\, z_1(r) = 2r\text{Even}\, g_1(r) + 2rg_2(-r^2),$$

$$x_3(r) + z_3(r) + x_2(r) - z_2(r) - 2\text{Even}\, x_1(r) - \frac{\text{Odd}\, z_1(r)}{r} =$$

$$c - d + 2g_2(r) - 2g_2(-r^2),$$

which determines the coefficients of $g_2(r) = \sum_{i=0}^{\infty} e_i r^i$ *splitting into parts as follows. Taking the odd part, we compute* e_i *with all odd* i, *the consider terms with powers* $4i$ *and* $4i+2$ *separately, find all* e_{4i+2} *and continue splitting into* parts. Having found $g_2(r)$, compute Even $g_1(r)$ from $(\divideontimes_2)$ and work out $h(r)$, Odd $g_1(r)$ from

$$\begin{cases} x_2(r) - z_2(r) - x_3(r) - z_3(r) = c + d + 2\mathrm{Odd}\, g_1(r) - 2h(r), \\ 2r\mathrm{Odd}\, x_1(r) + \mathrm{Even}\, z_1(r) = 2r\mathrm{Odd}\, g_1(r) + h(-r^2), \end{cases}$$

excluding Odd $g_1(r)$ and then splitting the result into parts as above.

Case x of another horizontal triple point $\not\wedge$ is similar to Case ix.

4.4.2 Bifurcation diagrams of codimension 2 singularities

The *bifurcation* diagram of a codimension 2 singularity δ from Definition 3.4 is formed by the pairs $(a, b) \in \mathbb{R}^2$ from the versal deformation of δ from Lemma 4.7. We will describe curves representing codimension 1 subspaces Σ_γ adjoined to Σ_δ in the space SL of all links $K \subset V$.

Oriented arcs in bifurcation diagrams of Fig. 4.8 are associated to canonical loops $\mathrm{CL}(K_{\pm\varepsilon}) \subset \mathrm{SL}$, where links $K_{\pm\varepsilon}$ are close to a given link K_0. At the zero critical moment, the loop $\mathrm{CL}(K_0)$ defines an arc through the origin $\{a = b = 0\}$. These arcs are transversal to the codimension 1 subspace $\Sigma^{(1)}$ apart from the cases below. In Figs. 4.8ix and 4.8x the canonical loop $\mathrm{CL}(K_s)$ is *parallel* to $\Sigma_{\wedge\!\vee}, \Sigma_{\wedge\wedge}, \Sigma_\curlyvee$ in the following sense: if $K \in \Sigma_{\wedge\!\vee}$, then $\mathrm{CL}(K) \subset \Sigma_{\wedge\!\vee} \cup \Sigma_{\searchleftarrow}$. If $K \in \Sigma_\curlyvee$, then $\mathrm{CL}(K) \subset \Sigma_\curlyvee \cup \Sigma_\curlywedge$. Similarly, $K \in \Sigma_{\wedge\wedge}$ implies that $\mathrm{CL}(K) \subset \Sigma_{\wedge\wedge} \cup \Sigma_{\wedge\!\wedge}$.

Lemma 4.8 *Figure 4.8 contains the bifurcation diagrams of the codimension 2 singularities* $\delta : \divideontimes, \divideontimes, \curlyvee, \curlyvee, \daleth, \divideontimes, \not\wedge, \bowtie, \prec, \wedge\!\wedge$ *and shows how the canonical loops* $\mathrm{CL}(K_{\pm\varepsilon})$ *intersect the adjoined codimension 1 subspaces* Σ_γ.

Proof 4.5 *In Cases i–v below the canonical loops transversally intersects all the singular subspaces since the tangents of intersecting arcs are not horizontal.*

Case i *of a quadruple point* $\divideontimes$. *There are 4 singular subspaces* $\Sigma_{\divideontimes}$ *intersecting each other transversally at the singular subspace* $\Sigma_{\divideontimes}$. *Using the normal form of* $\divideontimes$ *from Lemma 4.6, we show 4 subspaces in the bifurcation diagram of Fig. 4.8i, namely* $\{a = 0\}$ *(branches 1, 2, 4 intersect),* $\{b = 0\}$ *(branches 1, 3, 4 intersect),* $\{a = b\}$ *(branches 1, 2, 3 intersect),* $\{e(a + b) = b - a\}$ *(branches 2, 3, 4 intersect).*

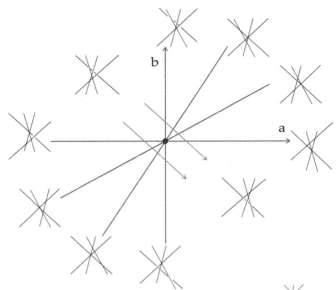

8i: the bifurcation diagram of a quadruple point

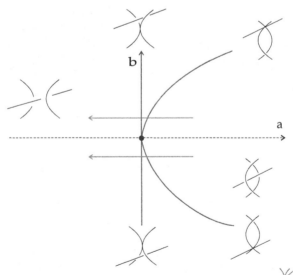

8ii: the bifurcation diagram of a tangent triple point

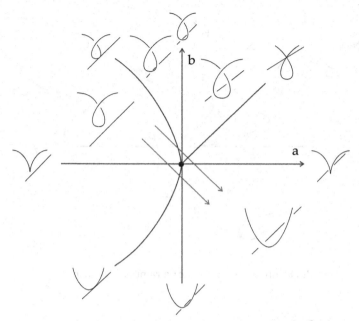

8iii: the bifurcation diagram of an intersected cusp

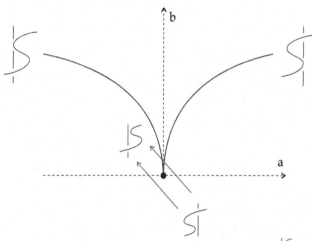

8iv: the bifurcation diagram of a cubic tangency

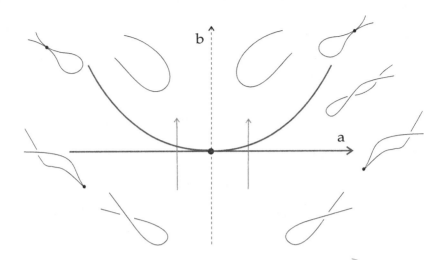

8v: the bifurcation diagram of a ramphoidal cusp

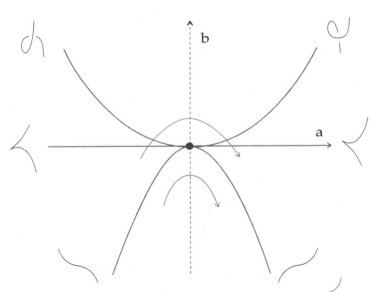

8vi: the bifurcation diagram of a horizontal cusp

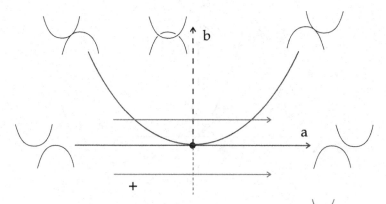

8vii : the bifurcation diagram of a mixed tangency

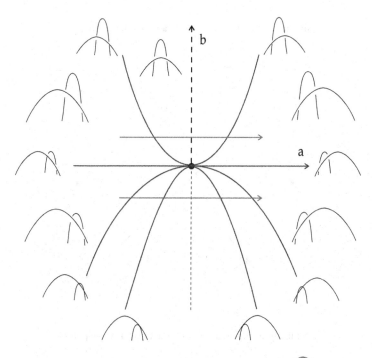

8viii : the bifurcation diagram of an extreme tangency

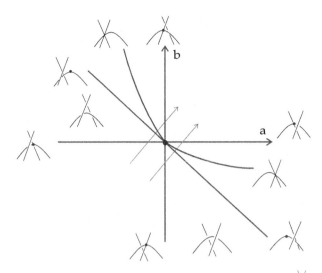

8ix: the bifurcation diagram of a horizontal triple point

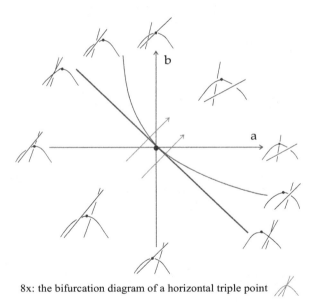

8x: the bifurcation diagram of a horizontal triple point

Figure 4.8: Bifurcation diagrams of codimension 2 singularities

Case ii *of a tangent triple point* �. *The branches* $\{x = z^2 - a\}$, $\{x = 0\}$ *have a tangency if* $a = 0$. *The triple point appears when* $z^2 - a = 0 = z - b$, *i.e.* $a = b^2$. *The bifurcation diagram of Fig. 4.8ii has 1 parabola and 1 line touching each other.*

Case iii *of an intersected cusp* ⅄. *The branch* $(r^3 - br, r^2)$ *has a self-intersection at* $r = \pm\sqrt{b}$, $b \geq 0$, *which becomes an ordinary cusp if* $b = 0$. *The self-intersection is a triple point when it is on the branch* $(r - a, r)$, *i.e.* $a = b$. *Finally, we get a simple tangency of* $(r^3 - br, r^2)$ *and* $(r - a, r)$ *if* $a = 2r^3 - r^2, b = 3r^2 - 2r$ *or* $3a - 2br = r^2$ *has a double root, i.e.* $b^2 + 3a = 0$. *The bifurcation diagram of Fig. 4.8iii contains 1 parabola, 1 line and 1 ray meeting at 0.*

Case iv *of a cubic tangency* ⌡. *The branch* $(r^3 - 3br + a, r)$ *has extrema of the x-coordinate at* $r = \pm\sqrt{b}$, *which lie on* $(0, r)$ *if* $r^3 - 3br + a = 0$, *i.e.* $a^2 = 4b^3$. *The only subspace* $\Sigma_{\curlyvee}$ *is adjoined to* $\Sigma_{\curlyvee}$ *in the bifurcation diagram of Fig. 4.8iv.*

Case v *of a ramphoidal cusp* ⌇. *The curve* $(r^5 + ar^3 + br, r^2)$ *has an ordinary cusp when* $\dot{x} = \dot{z} = 0$, *i.e.* $r = 0$ *and* $b = 0$, *and a self-tangency when* $5r^4 + 3ar^2 + b = 0$ *has two double roots, i.e.* $9a^2 = 20b$. *The bifurcation diagram of Fig. 4.8v contains 1 parabola and 1 line touching each other at 0.*

Case vi *of a horizontal cusp* ≺. *The curve* $(r^2, r^3 + ar^2 - br)$ *has a crossing at* $\pm r$, *hence* $r^3 = br$ *and* $r = \pm\sqrt{b}$, $b > 0$. *This crossing is critical, i.e.* $\dot{z} = 3r^2 + 2ar - b = 0$, *if* $b = a^2$. *The critical point becomes degenerate, i.e.* $\ddot{z} = 6r + 2a = 0$, *if* $b = -a^2/3$. *The subspace* $\Sigma_{\curlyvee}$ *of ordinary cusps, where* $\dot{x} = \dot{z} = 0$, *is represented by* $\{b = 0\}$. *The bifurcation diagram of Fig. 4.8vi shows 2 parabolas, 1 line and 1 ray meeting at 0. The arc associated to a canonical loop moves in the vertical direction and remains parallel to the parabola* $\{b = -a^2/3\}$ *representing the subspace* $\Sigma_{\curlyvee}$.

Case vii *of a mixed tangency* ⋈. *The branch* $(r, r^2 - b)$ *touches* $(r + a, -r^2)$ *if* $r^2 - b = -(r - a)^2$ *has a double root, i.e.* $a^2 = 2b$. *Both curves have extrema in the same horizontal line when* $b = 0$. *The bifurcation diagram of Fig. 4.8vii has 1 parabola and 1 line touching each other at 0.*

Case viii *of an extreme tangency* ⋀. *The branch* $(r, -2r^2 + b)$ *touches* $(r + a, -r^2)$ *if* $b - 2r^2 + b = -(r - a)^2$ *has a double root, i.e.* $2a^2 + b = 0$. *Both branches have extrema in the same horizontal line when* $b = 0$. *The branch* $(r, -2r^2 + b)$ *passes through an extremum of* $(r + a, -r^2)$ *at* $r = 0$ *if* $b = 2a^2$. *The branch* $(r + a, -r^2)$ *passes through an extremum of* $(r, -2r^2 + b)$ *at* $r = 0$

if $b = -a^2$. The bifurcation diagram of Fig. 4.8viii has 3 parabolas and 1 line touching each other at 0.

Case ix of a horizontal triple point ⋇. The branches $(r+a, r)$ and $(-r-b, r)$ pass through the extremum of $(r, -r^2)$ at $r = 0$ when $a = 0$ and $b = 0$, respectively. The crossing of $(r + a, r)$ and $(-r - b, r)$ at $r = -(a + b)/2$ lies in the same horizontal line with the extremum of $(r, -r^2)$ at $r = 0$ if $a + b = 0$. The branches $(r, -r^2)$, $(r + a, r)$ and $(-r - b, r)$ have a triple point if $r = -r^2 + a = r^2 - b$ or $(a - b)^2 = 2(a + b)$, which is a parabola in the bifurcation diagram of Fig. 4.8ix. The arc associated to a canonical loop is transversal to the subspaces, because only one tangent remains horizontal under the rotation.

Case x of another horizontal triple point ⋏ is similar to Case ix.

4.5   The diagram surface of a link

In this section the classification problem of generic links $K \subset V$ reduces to their diagram surfaces
DS(K) in the thickened torus $\mathbb{T} = A_{xz} \times S_t^1$, $A_{xz} = [-1, 1]_x \times S_z^1$.

4.5.1   The diagram surface of a link and generic surfaces

Briefly the diagram surface of a loop $\{K_t\}$ of links is the 1-parameter family of the diagrams $\mathrm{pr}_{xz}(K_t) \subset A_{xz} \times \{t\}$. This family can be considered as the union of link diagrams, i.e. as a 2-dimensional surface in the thickened torus $\mathbb{T} = A_{xz} \times S_t^1$.

Definition 4.14 Let $\{K_t\} \subset SL$ be a loop of links. The diagram surface DS($\{K_t\}$) $\subset A_{xz} \times S_t^1$ is formed by the diagrams $\mathrm{pr}_{xz}(K_t) \subset A_{xz} \times \{t\}$, $t \in S_t^1$. If K_t are knots, DS($\{K_t\}$) is the torus $S^1 \times S_t^1$ mapped to the thickened torus $\mathbb{T} = A_{xz} \times S_t^1$. The diagram surface DS(K) of an oriented link $K \subset V$ consists of the diagrams $\mathrm{pr}_{xz}(\mathrm{rot}_t(K)) \subset A_{xz} \times \{t\}$ and is oriented by the orientations of K and S_t^1.   ∎

Figure 4.9 shows vertical sections of DS(K) for a smoothed trefoil K from Fig. 4.2, $t \in [0, \pi]$. Each section is the diagram of a rotated trefoil $\mathrm{rot}_t(K)$ for some $t \in S_t^1$. Local extrema of $\mathrm{rot}_t(K)$ form horizontal circles parallel to

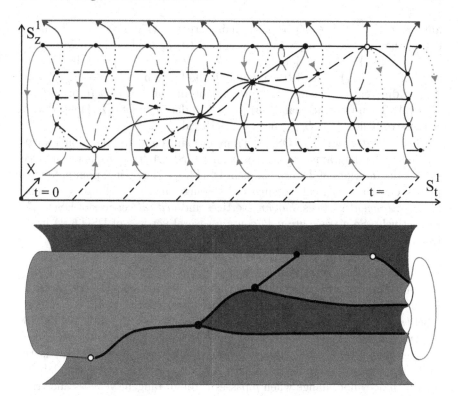

Figure 4.9: Half the diagram surface of a smoothed trefoil from Fig. 4.2

S_t^1. Several arcs in Fig. 4.9 are dashed or dotted, because they are invisible in the x-direction.

By Definition 4.12 the shift $t \mapsto t + \pi$ maps the surface $\mathrm{DS}(\{K_t\})$ to its image under the symmetry in S_z^1. Actually the link $K_{t+\pi}$ is obtained from K_t by the symmetry rot_π, i.e. the diagrams $\mathrm{pr}_{xz}(K_{t+\pi})$ and $\mathrm{pr}_{xz}(K_t)$ are symmetric for all $t \in S_t^1$. For a generic loop $\{K_t\}$, the vertical sections of $\mathrm{DS}(K)$ are the diagrams $\mathrm{pr}_{xz}(K_t)$ and allow the codimension 1 singularities $\times, \times, \curlyvee, \curlywedge$ only. It follows from the fact that any critical point of $\mathrm{pr}_z : K_t \to S_z^1$ remains critical under rot_t.

For any $t \in S_t^1$, the points from $K_t \cap (\mathrm{D}_{xy} \times \{z = \pm 1\})$ and the critical points of $\mathrm{pr}_z : K_t \to S_z^1$ divide the i-th component of K_t into arcs $A_{t,i,q}$, $q = 1, \ldots, n_i$. The total number of these arcs does not depend on t since any critical point $a_t \in K_t$ of pr_z remains critical while t varies. The union $\cup a_t$ of the extrema of $\mathrm{pr}_z : K_t \to S_z^1$ for all $t \in S_t^1$ splits into *critical circles* C_i of $\mathrm{DS}(\{K_t\})$. The union $B_{i,q} = \cup A_{t,i,q}$ over all $t \in S_t^1$ is called a *trace band* of $\mathrm{DS}(\{K_t\})$. The 3 trace bands in the bottom picture of Fig. 4.9 are dashed differently. The arcs $A_{t,i,q}$ are monotonic with respect to $\mathrm{pr}_{zt} : K_t \to S_z^1 \times \{t\}$. Then the trace bands project 1-1 under $\mathrm{pr}_{zt} : \mathrm{DS}(\{K_t\}) \to S_z^1 \times S_t^1$. Successive bands $B_{i,q}$, $B_{i+1,q}$ meet at a critical circle.

The *singular* points of $\mathrm{DS}(\{K_t\})$ are crossings and codimension 1 singularities of the diagrams $\mathrm{pr}_{xz}(K_t)$ over all $t \in S_t^1$. A *trace arc* is an intersection of the interiors of 2 trace bands in $\mathrm{DS}(\{K_t\})$. The triple points, tangent points, cusps and critical crossings of link diagrams $\mathrm{pr}_{xz}(K_t)$ are called *triple vertices*, *tangent* vertices, *hanging* vertices and *critical* vertices of $\mathrm{DS}(\{K_t\})$, respectively. So a trace arc may contain several vertices of $\mathrm{DS}(\{K_t\})$ in the usual sense.

Take a singular point $p \in \mathrm{DS}(K)$ that is not a vertex and does not belong to a critical circle of $\mathrm{DS}(K)$. Then p is a double crossing of two arcs $A_{t,i,q}$ and $A_{t,j,s}$ in a diagram $\mathrm{pr}_{xz}(K_t)$. If the arc $A_{t,i,q}$ passes over (respectively, under) $A_{t,j,s}$ then associate to p the *label* $(q_i s_j)$ (respectively, the *reversed* label $(s_j q_i)$). If K_t is a knot then we miss the indices $i, j = 1$ as in Fig. 4.3.

Trace arcs of $\mathrm{DS}(\{K_t\})$ end at hanging vertices, meet each other at critical vertices and intersect at triple vertices. Each trace arc of $\mathrm{DS}(K)$ is the evolution trace of a double crossing in $\mathrm{A}_{xz} \times S_t^1$ while t varies. The label of a point p does not change when p passes through tangent vertices and triple vertices.

The diagram surface can be defined for any loop of links and can be extremely complicated. The surfaces corresponding to generic loops are simple and play the role of general link diagrams in dimension 3. As in the case of links, we define a generic surface associated to a generic loop. A generic surface will be an immersed surface with all combinatorial features of diagram surfaces of generic loops. For any generic surface, a corresponding generic loop is constructed in Lemma 4.11.

Definition 4.15 *Decompose S_i^1 into arcs $A_{i,1}, \ldots, A_{i,n_i}$. Introduce the* trace *bands $B_{i,q} = A_{i,q} \times S_t^1$, $q = 1, \ldots, n_i$. A generic* surface S is the image of *a smooth map $h : (\sqcup_{i=1}^m S_i^1) \times S_t^1 = \cup_{i=1}^m (\cup_{q=1}^{n_i} B_{iq}) \to \mathrm{A}_{xz} \times S_t^1$ such that Conditions (i)–(v) hold*

(i) Conditions *on* symmetry *and* trace bands.

- under $t \mapsto t + \pi$ the surface S maps to its image under the symmetry in S_z^1;

- each trace band $B_{i,q} \subset S$ projects one-to-one under $\mathrm{pr}_{zt} : S \to S_z^1 \times S_t^1$.

The surface S should be simple enough. More formally we require the following.

(ii) Conditions on sections $D_t = S \cap (A_{xz} \times \{t\})$, $t \in S_t^1$.

There are finitely many critical moments $t_1, \ldots, t_l \in S_t^1$ such that

- for all $t \notin \{t_1, \ldots, t_l\}$, the sections $\{D_t\}$ are general diagrams;

- for each $t = t_1, \ldots, t_l$, the section D_t has one of the singularities $\divideontimes, \times, \curlyvee, \curlywedge$;

- while t passes a critical moment, D_t changes by a move I–IV in Fig. 4.5.

Conditions (ii) on sections imply some restrictions on trace bands. These requirements can be stated independently to define trace arcs and critical circles.

(iii) Conditions on trace arcs and critical circles:

- a trace arc is an intersection of the interiors of 2 trace bands $B_{i,q}$ and $B_{j,s}$;

- a critical circle $C_{i,q}$ is the common boundary of successive bands $B_{i,q}$, $B_{i,q+1}$.

The arcs defined above allow us to introduce vertices of a generic surface S.

(iv) Conditions on vertices:

- a triple vertex is a transversal intersection of 3 trace bands B_{iq}, B_{js}, B_{kr};

- a hanging vertex of S is the endpoint of a trace arc in $B_{i,q} \cap B_{i,q+1}$;

- a critical vertex is the intersection of a critical circle $C_{i,q}$ and $B_{j,s} \not\supset C_{i,q}$;

- *a* tangent *vertex is a critical point of* pr_t *on the interior of a trace arc;*

- *all the* vertices *are distinct and map on different points under* $\mathrm{pr}_t : S \to S_t^1$.

Finally fix labels (i, q) and (j, s). Take a trace arc from the intersection $B_{i,q} \cap B_{j,s}$ of interiors of 2 trace bands. Endow the chosen arc with a label: either $(q_i s_j)$ or $(s_j q_i)$ in such a way that the following restrictions apply.

(v) Conditions *on* labels*:*

- *under the time shift $t \mapsto t + \pi$, each label reverses: $(q_i s_j) \mapsto (s_j q_i)$;*

- *trace arcs intersecting at a triple vertex are endowed with $(q_i s_j)$, $(s_j r_k)$, $(q_i r_k)$;*

- *a hanging vertex is endowed with the label of the trace arc containing it;*

- *each circle $C_{i,q}$ has 2 hanging vertices endowed with $((q+1)_i, q_i)$, $(q_i, (q+1)_i)$;*

- *if a trace band $B_{j,s}$ intersects a critical circle $C_{i,q}$ in a vertex c then the label at c transforms as follows: $(q_i s_j) \leftrightarrow ((q+1)_i, s_j)$ or $(s_j q_i) \leftrightarrow (s_j, (q+1)_i)$.* ∎

To get the following result compare Definitions 4.12, 4.14 with Definition 4.15.

Lemma 4.9 *For any generic loop L of links, the diagram surface $\mathrm{DS}(L)$ is a generic surface in the sense of Definition 4.15.*

4.5.2   Three-dimensional moves on generic surfaces

Definition 4.16 *A smooth family of surfaces $\{S_r \subset \mathrm{A}_{xz} \times S_t^1\}$, $r \in [0, 1]$, is an* equivalence *if there are finitely many critical moments $r_1, \ldots, r_k \in [0, 1]$ such that*

- *for all non-critical moments $r \notin \{r_1, \ldots, r_k\}$, the surfaces S_r are generic;*

- *if r passes through a critical moment, S_r changes by a move in Fig. 4.10.* ∎

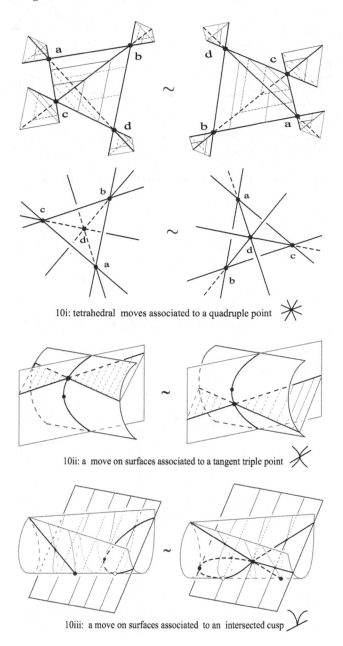

10i: tetrahedral moves associated to a quadruple point

10ii: a move on surfaces associated to a tangent triple point

10iii: a move on surfaces associated to an intersected cusp

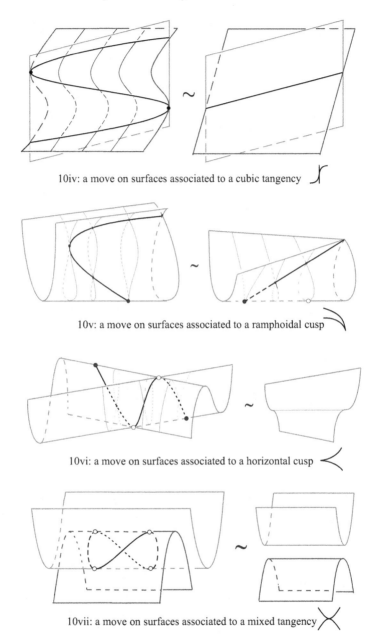

10iv: a move on surfaces associated to a cubic tangency

10v: a move on surfaces associated to a ramphoidal cusp

10vi: a move on surfaces associated to a horizontal cusp

10vii: a move on surfaces associated to a mixed tangency

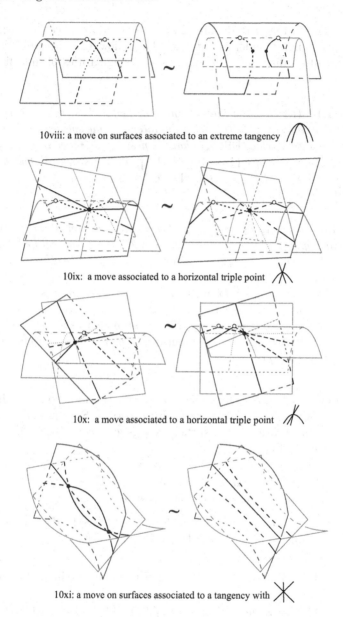

Figure 4.10: Three-dimensional moves on diagram surfaces

Each move in Fig. 4.10 denotes 2 symmetric moves since the surfaces S_r are symmetric in S_z^1 under $t \mapsto t + \pi$. The following claim will be proved using bifurcation diagrams of codimension 2 singularities of link diagrams, see Lemma 4.8.

Lemma 4.10 (a) *Suppose that a family of loops $\{L_s\}$, $s \in [-1,1]$, in the space* SL *of all links $K \subset V$ transversally intersects the subspace $\Sigma^{(2)}$ at $s = 0$. Then the diagram surface* DS(L_s) *changes near 0 by a move in Figs. 4.10i–x.*
(b) *If a family of loops $\{L_s\}$, $s \in [-1,1]$, in the space* SL *has a simple tangency with $\Sigma_{\times}$ at $s = 0$, then* DS(L_s) *changes near 0 by the move in Fig. 4.10xi.*

Sketch 4.4 *The pictures in Fig. 4.10 are obtained from the corresponding pictures in Fig. 4.8. For instance, in Fig. 4.8iii the canonical loop* CL($K_{-\varepsilon}$) *meets 3 subspaces $\Sigma_{\curlyvee}, \Sigma_{\times}, \Sigma_{\curlywedge}$. Therefore the surface* DS($K_{-\varepsilon}$) *has three distinguished points: a hanging vertex, a tangent vertex and a critical one as in Fig. 4.10iii. Right after the move when all three points pass through each other, the surface* DS($K_{+\varepsilon}$) *has 4 interesting points: three have the previous types, the new one is a triple vertex. This situation agrees with 4 intersections of* CL($K_{+\varepsilon}$) *with codimension 1 subspaces in Fig. 4.8iii. The remaining cases are absolutely analogous.*

We produced Fig. 4.10 first using our geometric intuition and then justified the moves applying the singularity theory in Section 4.4. Since the family of sections in a generic surface is a general equivalence of diagrams then Lemma 4.11 follows.

Lemma 4.11 (a) *For any generic surface S, there is a generic loop L of links such that the diagram surface* DS(L) *coincides with S.*

(b) *For any equivalence of surfaces $\{S_r \subset A_{xz} \times S_t^1\}$, there is a generic homotopy of loops $\{L_r\}$ such that* DS(L_r) $= S_r$, $r \in [0,1]$.

Lemma 4.10 and Definition 4.13 of a generic homotopy imply Lemma 4.12.

Lemma 4.12 *Any generic homotopy of loops $\{L_s\}$, $s \in [0,1]$ in the space* SL *provides an equivalence $\{$*DS(L_s)$\}$ *of diagram surfaces.*

Lemma 4.13 *Let L_0, L_1 be generic loops of links. If $\mathrm{DS}(L_0)$ and $\mathrm{DS}(L_1)$ are equivalent in the sense of Definition 4.16, then L_0 and L_1 are generically homotopic.*

Proof 4.6 *Any equivalence of diagram surfaces gives rise to a smooth family of loops $\{L_r\}$ by Lemma 4.11b. The constructed family $\{L_r\}$ is a generic homotopy since all moves in Fig. 4.10 correspond to singularities in the sense of Definition 4.10.*

By Lemmas 4.12 and 4.13 the classification of generic links reduces to the equivalence problem for their diagram surfaces.

Proposition 4.5 *Generic links K_0, K_1 are generically equivalent if and only if the diagram surfaces $\mathrm{DS}(K_0), \mathrm{DS}(K_1)$ are equivalent in the sense of Definition 4.16.*

The isotopy class of a link can be easily reconstructed from its plane diagram, hence from its diagram surface with labels. Formally, one has the following.

Lemma 4.14 *Suppose that the diagram surface $\mathrm{DS}(K)$ of a generic link K is given, but K is unknown. Then one can reconstruct the isotopy class of $K \subset V$.*

4.6   The trace graph of a link as a link invariant

4.6.1   The trace graph of a link and generic trace graphs

Here the classification of links $K \subset V$ will be reduced to their trace graphs.

Definition 4.17 *Let $S \subset \mathrm{A}_{xz} \times S_t^1$ be the diagram surface of a loop of links. The trace graph $\mathrm{TG}(S)$ is the self-intersection of S, i.e. a finite graph embedded into $\mathrm{A}_{xz} \times S_t^1$. The trace graph $\mathrm{TG}(K)$ of a link K is the trace graph of its diagram surface $\mathrm{DS}(K)$. The trace arcs of $\mathrm{DS}(K)$ are called trace arcs of $\mathrm{TG}(K)$. The trace graph inherits the vertices and labels from $\mathrm{DS}(K)$.* ∎

Definition 4.18 *A finite graph $G \subset A_{xz} \times S_t^1$ is generic if Conditions (i)–(ii) hold.*

(i) Conditions *on* trace arcs *and* vertices.

- *the graph G consists of finitely many* trace arcs, *which are monotonic arcs with respect to the orthogonal projection* $\mathrm{pr}_z : G \to S_z^1$;

- *any endpoint of a trace arc of G has either degree 1 (a* hanging vertex $\leftarrow$ *) or degree 2 (a* critical vertex $\multimap$ *);*

- *the critical vertices of G coincide with the critical points of* $\mathrm{pr}_z : G \to S_z^1$;

- *trace arcs of G intersect transversally at* triple vertices $(\!\times\!)$;

- *the critical points of* $\mathrm{pr}_z : G \to S_t^1$ *are called* tangent vertices $(\subset)$.

(ii) Conditions *on* labels.

- *each trace arc of G is labelled with a label $(q_i s_j)$ as in Definition 4.2;*

- *under $t \mapsto t + \pi$ the graph G maps to its image under the symmetry in S_z^1;*

- *under the time shift $t \mapsto t + \pi$ each label $(q_i s_j)$ reverses to $(s_j q_i)$;*

- *every triple vertex $v \in G$ is labelled with a triplet $(q_i s_j)$, $(s_j r_k)$, $(q_i r_k)$ consisting of the labels associated to the trace arcs passing through v;*

- *each hanging vertex is labelled with the label of the corresponding trace arc;*

- *for any i and $q = 1, \ldots, n_i$, there are exactly two hanging vertices of G labelled with $((q+1)_i, q_i)$ and $(q_i, (q+1)_i)$, respectively;*

- *at every critical vertex of G the labels of trace arcs may transform as follows:*
  either $(q_i s_j) \leftrightarrow (q_i, (s \pm 1)_j)$ or $(q_i s_j) \leftrightarrow ((q \pm 1)_i, s_j)$.   ∎

A trace arc of a generic graph may consist of several edges in the usual sense.

Lemma 4.15 (a) *For any generic surface S, the trace graph $\mathrm{TG}(S)$ is generic in the sense of Definition 4.18. So the trace graph $\mathrm{TG}(K)$ of a generic link K is generic.*

Proof 4.7 *Conditions (i)–(v) of Definition 4.15 imply Conditions (i)–(ii) of Definition 4.18.*

Definition 4.19 *A smooth family of trace graphs $\{G_s\}$, $s \in [0,1]$, is called an* equivalence *if there are finitely many critical moments $s_1, \ldots, s_k \in [0,1]$ such that*

- *for all non-critical moments $s \notin \{s_1, \ldots, s_k\}$, the trace graphs G_s are generic;*

- *if s passes through a critical moment, G_s changes by a move in Fig. 4.11.*

∎

The moves in Fig. 4.11 should be considered locally, i.e. the diagrams do not change outside the pictures. Various mirror images of the moves are also possible. Moreover, some labels $s + 1$ can be replaced by $s - 1$ and vice versa. Trace graphs are symmetric under $t \mapsto t + \pi$, i.e. each move in Fig. 4.11 denotes two symmetric moves. The most non-trivial moves are *tetrahedral* moves Fig. 4.11i and *trihedral* moves Fig. 4.11xi. Their geometric interpretation at the level of links is shown in Fig. 4.12.

Notice that both moves in Fig. 4.11i can be realized for links and closed braids. In general a tetrahedral move corresponds to a link or a braid with a horizontal quadrisecant. Geometrically two arcs intersect a wide band bounded by another two arcs. Under a tetrahedral move, the two intersection points swap their heights as in Fig. 4.12. The first picture of Fig. 4.11i applies when the intermediate oriented arcs go together from one side of the band to another like $\rightrightarrows$. The second picture means that the arcs are antiparallel as in the British rail mark $\rightleftarrows$. It is easier to understand Lemma 4.16 first for knots, when the indices $i, j = 1$ can be missed.

Lemma 4.16 (a) *For a generic trace graph G such that $G \cap (\mathrm{A}_{xz} \times \{0\})$ are crossings of a general diagram, there is a generic surface S such that $\mathrm{TG}(S) = G$.*

(b) *For any equivalence of trace graphs $\{G_r\}$, there is an equivalence of surfaces S_r with $\mathrm{TG}(S_r) = G_r$, $r \in [0,1]$.*

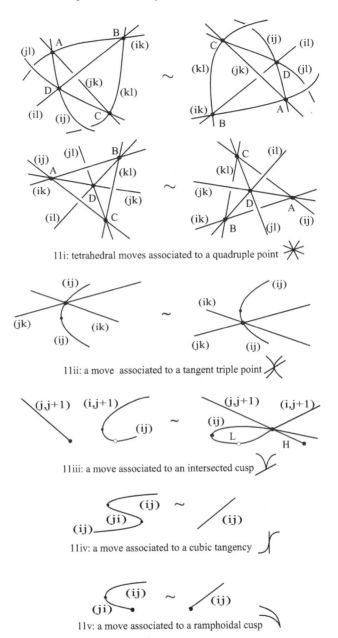

11i: tetrahedral moves associated to a quadruple point

11ii: a move associated to a tangent triple point

11iii: a move associated to an intersected cusp

11iv: a move associated to a cubic tangency

11v: a move associated to a ramphoidal cusp

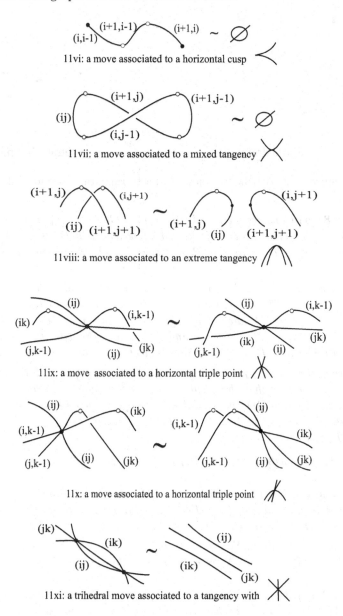

Figure 4.11: Moves on trace graphs

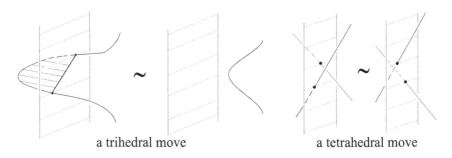

a trihedral move   a tetrahedral move

Figure 4.12: A trihedral move and a tetrahedral move for links

Proof 4.8 (a) *Consider a vertical section* $P_t = G \cap (A_{xz} \times \{t\})$ *not containing vertices of* G. *Then* P_t *is a finite set of points with labels* $(q_i s_j)$, *where* $i, j \in \{1, \dots, m\}$. *The points in* P_t *will play the role of crossings of sections of* S.

The labelled set P_t *defines the Gauss diagram* GD_t *as follows, see Definition 4.7. Take* $\sqcup_{i=1}^{m} S_i^1$, *split each circle* S_i^1 *into* n_i *arcs and number them by* $1, \dots, n_i$ *according to the orientation. We mark several points in the q-th arc of* S_i^1 *in a 1-1 correspondence and the same order with the points of* P_t *projected under* $\mathrm{pr}_z : P_t \to S_z^1$ *and having labels* $(q_i s_j)$ *or* $(s_j q_i)$, $s = 1, \dots, n_j$.

So each point of P_t *gives 2 marked points in* $\sqcup_{i=1}^{m} S_i^1$, *labelled with* $(q_i s_j)$ *and* $(s_j q_i)$. *Connect them by a chord and get the Gauss diagram* GD_t. *The zero Gauss diagram* GD_0 *is realizable by the given general diagram. Hence all Gauss diagrams* GD_t *give rise to a family of diagrams* D_t, *i.e. to a surface* $S = \cup (D_t \times \{t\})$.

(b) *Apply the construction from* (a) *to each trace graph* G_r, $r \in [0, 1]$.

Proposition 4.6 (a) *Trace graphs* $TG(S_0), TG(S_1)$ *of generic surfaces are equivalent in the sense of Definition 4.19 if and only if the surfaces* S_0, S_1 *are equivalent in the sense of Definition 4.16.*

(b) *Generic surfaces* S_0, S_1 *are equivalent in the sense of Definition 4.16 if and only if* $TG(S_0), TG(S_1)$ *are equivalent in the sense of Definition 4.19.*

Proof 4.9 (a), (b) *Any equivalence* $\{S_r\}$ *of surfaces gives rise to the equivalence* $TG(S_r)$ *of trace graphs. Any equivalence of trace graphs gives rise to a smooth family of diagram surfaces* $\{S_r\}$ *by Lemma 4.16b. The family*

$\{S_r\}$ is an equivalence of diagram surfaces since the moves in Fig. 4.11 are restrictions of the moves in Fig. 4.10.

Theorem 4.1 directly follows from Propositions 4.2, 4.3, 4.5 and 4.6.

Lemma 4.17 Suppose that the trace graph $G = \mathrm{TG}(K)$ of a generic link K is given, but K is unknown. Then one can construct a generic link K' equivalent to K.

Proof 4.10 Lemma 4.16a provides a generic surface S such that $\mathrm{TG}(S) = G$. Due to labels of trace arcs, the section $D_0 = S \cap (\mathrm{A}_{xz} \times \{0\})$ gives rise to a link $K' \subset V$ with $\mathrm{pr}_{xz}(K') = D_0$. The link K' can be assumed to be generic by Proposition 4.2a and is equivalent to K since K and K' have the same Gauss diagram.

4.6.2 Combinatorial construction of a trace graph

Lemma 4.18 Let $K \subset V$ be a link with $2e$ extrema of the projection pr_z : $K \to S_z^1$ and l crossings in the diagram $\mathrm{pr}_{xz}(K)$. Let the extrema and intersection points from $K \cap (\mathrm{D}_{xy} \times \{z = \pm 1\})$ divide K into n arcs monotonic with respect to pr_z. Then K is isotopic in V to a link K' such that $\mathrm{TG}(K')$ contains $2l(n-2)$ triple vertices, $4(n-e-1)e$ critical vertices and $2e$ hanging vertices.

Proof 4.11 Take a generic link K' smoothly equivalent to K and having an isotopic plane diagram. We split K' by horizontal planes into several horizontal slices such that each slice contains exactly one crossing or one extremum with respect to $\mathrm{pr}_z : K' \to S_z^1$. We may assume that all maxima are above all minima, otherwise deform K' accordingly. To each slice we associate the corresponding elementary trace graph and glue them together, see examples in Fig. 4.13 and Fig. 4.14.

Figure 4.13 shows two explicit examples for the opposite crossings in the braid group B_4. In general we mark out the points $\psi_k = 2^{1-k}\pi$, $k = 0, \ldots, n-1$ on the boundary of the bases $\mathrm{D}_{xy} \times \{\pm 1\}$. The 0-th point $\psi_0 = 2\pi$ is the n-th point.

The crucial feature of the distribution $\{\psi_k\}$ is that all straight lines passing through two points ψ_j, ψ_k are not parallel to each other. Firstly we draw all strands in the cylinder $\partial \mathrm{D}_{xy} \times [-1,1]_z$. Secondly we approximate with the

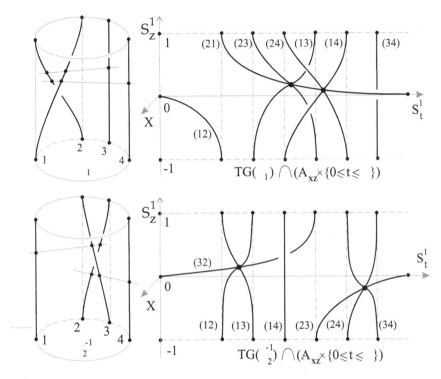

Figure 4.13: Half trace graphs of the 4-braids $\sigma_1, \sigma_2^{-1} \in B_4$

first derivative the strands forming a crossing by smooth arcs, see the left pictures in Fig. 4.13.

Then each elementary braid σ_i constructed as above has exactly $n-2$ horizontal trisecants through the strands $i, i+1$ and j for $j \neq i, i+1$. Each trisecant is associated to a triple vertex of the trace graph, see 4 horizontal trisecants in the left picture of Fig. 4.13. The trace graphs in Fig. 4.13 are not generic in the sense of Definition 4.18, e.g. parallel strands 3 and 4 lead to the vertical trace arc labelled with (34). But we may slightly deform such a trace graph to make it generic.

In the first picture of Fig. 4.14 the arc with a maximum is the intersection of the cylinder $\partial D_{xy} \times [-1, 1]_z$ with an inclined plane containing the straight line 1-2 in the base $D_{xy} \times \{-1\}$. The highest maximum of K' leads to exactly $2(n-2e)$ critical vertices (with symmetric images under $t \mapsto t+\pi$), the next

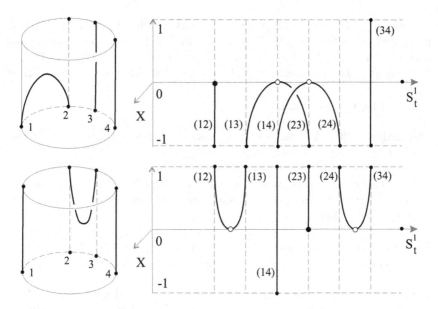

Figure 4.14: Half trace graphs of elementary blocks containing extrema

maximum gives $2(n-2e+2)$ critical vertices and so on, i.e. the total number is $2(n-2e) + 2(n-2e+2) + \cdots + 2(n-2) = 2(n-e-1)e$. The number of critical vertices associated to minima of K' is the same. Moreover each of $2e$ extrema gives one hanging vertex.

4.7 1-parameter knot theory for knots which are closed braids in the solid torus

For the convenience of the reader we repeat (and refine) in this section our main constructions but only in the special case of those knots in the solid torus which are closed braids.

4.7.1 The canonical loop for closed braids

We identify $\mathbb{R}^3 \setminus z - axes$ with the standard solid torus $V = S^1 \times D^2 \hookrightarrow \mathbb{R}^3 \setminus z - axes$. We identify the core of V with the unit circle in $\mathbb{R}^2$.

Let $rot(V)$ denote the S^1-parameter family of diffeomorphisms of V which is defined in the following way: we rotate the solid torus monotonously and with constant speed around its core by the angle t, $t \in [0, 2\pi]$, i.e. all discs $(\phi = const) \times D^2$ stay invariant and are rotated simultaneously around their center.

Let $\hat{\beta}$ be a closed braid.

Definition 4.20 *The* canonical loop $rot(\hat{\beta}) \in M(\hat{\beta})$ *is the oriented loop induced by* $rot(V)$.

The canonical loop is an analogue of Gramain's loop for long knots.

The following lemma (see [20]) is an immediate corollary of the definition of the canonical loop.

Lemma 4.19 *Let* $\hat{\beta}_s, s \in [0, 1]$, *be an isotopy of closed braids in the solid torus. Then* $rot(\hat{\beta}_s), s \in [0, 1]$, *is a homotopy of loops in* $M(\hat{\beta})$.

Evidently, the canonical loop can be defined for an arbitrary link in V in exactly the same way. However, in the case of closed braids we can give an another (combinatorial) definition, which makes concrete calculations much easier.

Let $\Delta \in B_n$ be Garside's element, i.e.

$$\Delta = (\sigma_1 \sigma_2 \ldots \sigma_{n-1})(\sigma_1 \sigma_2 \ldots \sigma_{n-2}) \ldots (\sigma_1 \sigma_2)(\sigma_1).$$

Its square Δ^2 is a generator of the center of B_n (see e.g. [5]). Geometrically, Δ^2 is the full twist of the n strings.

Definition 4.21 *Let* $\gamma \in B_n$ *be a braid with closure isotopic to* $\hat{\beta}$. *Then the* combinatorial canonical loop $rot(\gamma)$ *is defined by the following sequence of braids:*

$$\gamma \to \Delta\Delta^{-1}\gamma \to \Delta^{-1}\gamma\Delta \to \cdots \to \Delta^{-1}\Delta\gamma' \to \gamma' \to \Delta\Delta^{-1}\gamma' \to \Delta^{-1}\gamma'\Delta \to$$
$$\cdots \to \Delta^{-1}\Delta\gamma \to \gamma$$

Here, the first arrow consists only of Reidemeister II moves, the second arrow is a cyclic permutation of the braid word (which corresponds to an isotopy of the braid diagram in the solid torus) and the following arrows consist of "pushing Δ monotonously from the right to the left through the braid γ". We obtain a braid γ', which is just γ with each generator σ_i replaced by σ_{n-i}, and we start again.

We give below a precise definition in the case $n = 3$. The general case is a straightforward generalization which is left to the reader. $\Delta = \sigma_1\sigma_2\sigma_1$ for $n = 3$. We have just to consider the following four cases:

$$\sigma_1\Delta = \sigma_1(\sigma_1\sigma_2\sigma_1) \to \sigma_1(\sigma_2\sigma_1\sigma_2) = \Delta\sigma_2$$
$$\sigma_2\Delta = (\sigma_2\sigma_1\sigma_2)\sigma_1 \to (\sigma_1\sigma_2\sigma_1)\sigma_1 = \Delta\sigma_1$$
$$\sigma_1^{-1}\Delta = \sigma_1^{-1}(\sigma_1\sigma_2\sigma_1) \to \sigma_2\sigma_1 \to (\sigma_1\sigma_1^{-1})\sigma_2\sigma_1 \to \sigma_1(\sigma_2\sigma_1\sigma_2^{-1}) = \Delta\sigma_2^{-1}$$
$$\sigma_2^{-1}\Delta = \sigma_2^{-1}(\sigma_1\sigma_2\sigma_1) \to (\sigma_1\sigma_2\sigma_1^{-1})\sigma_1 \to \sigma_1\sigma_2 \to \sigma_1\sigma_2\sigma_1\sigma_1^{-1} = \Delta\sigma_1^{-1}.$$

Notice, that the sequence is canonical in the case of a generator and almost canonical in the case of an inverse generator. Indeed, we could replace the above sequence $\sigma_1^{-1}\Delta \to \Delta\sigma_2^{-1}$ by

$$\sigma_1^{-1}(\sigma_1\sigma_2\sigma_1) \to \sigma_2\sigma_1 \to \sigma_2\sigma_1\sigma_2\sigma_2^{-1} \to (\sigma_1\sigma_2\sigma_1)\sigma_2^{-1}.$$

But it turns out that the corresponding canonical loops in $M(\hat{\beta})$ differ just by a homotopy which passes once transversely through a stratum of $\Sigma^{(2)}_{trans-self}$ and our one-cocycles are invariant under this homotopy.

Let c be the word length of γ. Then we use exactly $2c(n-2)$ braid relations (or equivalently, Reidemeister III moves) in the combinatorial canonical loop. This means that the corresponding loop in $M(\hat{\beta})$ intersects $\Sigma^{(1)}_{tri}$ transversely in exactly $2c(n-2)$ points.

One easily sees that the combinatorial canonical loop $rot(\gamma)$ from Definition 4.21 differs from the geometrical canonical loop $rot(\hat{\beta})$ from Definition 4.20 only by loops which correspond to rotations of the solid torus along its core (i.e. around the axis of the complementary solid torus in S^3). But each such loop is just an isotopy of diagrams with respect to pr and does not intersect the discriminant $\Sigma^{(i)}, i > 0$, at all.

4.7.2 The trace graph of the canonical loop for closed braids

The trace graph $TL(rot(\hat{\beta}))$ is our main combinatorial object, compare [20] and [21]. It is an oriented singular link in a thickened torus. All its singularities are ordinary triple points. Let $\hat{\beta}_t, t \in S^1$, be the (oriented) family of closed braids corresponding to the canonical loop $rot(\hat{\beta})$. We assume that the loop $rot(\hat{\beta})$ is a generic loop. Let $\{p_1^{(t)}, p_2^{(t)}, \ldots, p_m^{(t)}\}$ be the set of double points of $pr(\hat{\beta}_t) \subset S^1_\phi \times \mathbb{R}_\rho$. The union of all these crossings for all $t \in S^1$

forms a link $TL(rot(\hat{\beta})) \subset (S_\phi^1 \times \mathbb{R}_\rho^+) \times S_t^1$ (i.e. we forget the coordinate $z(p_i^{(t)})$). $TL(rot(\hat{\beta}))$ is non-singular besides ordinary triple points which correspond exactly to the triple points in the family $pr(\hat{\beta}_t)$. A generic point of $TL(rot(\hat{\beta}))$ corresponds just to an ordinary crossing $p_i^{(t)}$ of some closed braid $\hat{\beta}_t$. Let $t : TL(rot(\hat{\beta})) \to S_t^1$ be the natural projection. We orient the set of all generic points in $TL(rot(\hat{\beta}))$ (which is a disjoint union of embedded arcs) in such a way that the local mapping degree of t at $p_i^{(t)}$ is $+1$ if and only if $p_i^{(t)}$ is a positive crossing (i.e. it corresponds to a generator of B_n, or equivalently, its *writhe* $w(p_i^{(t)}) = +1$).

The arcs of generic points come together in the triple points and in points corresponding to an ordinary self-tangency in some $pr(\hat{\beta})$. But one easily sees that the above defined orientations fit together to define an orientation on the natural resolution $T\tilde{L}(rot(\hat{\beta}))$ of $TL(rot(\hat{\beta}))$, compare also [16]. $T\tilde{L}(rot(\hat{\beta}))$ is a union of oriented circles, called *trace circles*. We can attach *stickers* $i \in \{1, 2, \dots, n-1\}$ to the edges of $T\tilde{L}(rot(\hat{\beta}))$ in the following way: each edge of $T\tilde{L}(rot(\hat{\beta}))$ corresponds to a letter in a braid word. Indeed, each generic point in an edge corresponds to an ordinary crossing of a braid projection and, hence, to some σ_i or some σ_i^{-1}. We attache to this edge the number i. The information about the exponent $+1$ or -1 is contained in the orientation of the edge.

We identify $H_1(V)$ with $\mathbb{Z}$ by sending the homology class which is generated by the closed 1-braid to the generator $+1$. If $\hat{\beta}$ is a knot then we can attache to each trace circle a *homological marking* $a \in H_1(V)$ in the following way: let p be a crossing corresponding to a generic point in the trace circle. We smooth p with respect to the orientation of the closed braid. The result is an oriented 2-component link. The component of this link which contains the under-cross which goes to the over-cross at p is called p^+. We associate now to p the homology class $a = [p^+] \in H_1(V)$. One easily sees that $a \in \{1, 2, \dots, n-1\}$ and that the class a does not depend on the choice of the generic point in the trace circle. Indeed, the two crossings involved in a Reidemeister II move have the same homological marking and a Reidemeister III move does not change the homological marking of any of the three involved crossings.

We could see $TL(rot(\hat{\beta}))$ as an object which takes into account simultaneously all possible directions of projections of $\hat{\beta}$ into isotopic properly embedded annuli in V which contain the core of V.

Figure 4.15: A trihedron

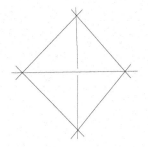

Figure 4.16: A tetrahedron

Evidently, crossings with different homological markings belong to different trace circles. Surprisingly, the inverse is also true, see Lemma 3.1 in [21].

Lemma 4.20 *Let* $TL(rot(\hat{\beta}))$ *be the trace graph of the closure of a braid* $\beta \in B_n$, *such that* $\hat{\beta}$ *is a knot. Then* $TL(rot(\hat{\beta}))$ *splits into exactly n-1 trace circles. They have pairwise different homological markings.*

Consequently, the trace circles are characterized by their homological markings. Notice, that the set of homological markings is independent of the word length of the braid (which has a knot as closure).

4.7.3 A higher order Reidemeister theorem for trace graphs of closed braids

Definition 4.22 *A trihedron is a 1-dimensional subcomplex of* $TL(rot(\hat{\beta}))$ *which is contractible in the thickened torus and which has the form as shown in Fig. 4.15.*

Definition 4.23 *A tetrahedron is a 1-dimensional subcomplex of* $TL(rot(\hat{\beta}))$ *which is contractible in the thickened torus and which has the form as shown in Fig. 4.16.*

Figure 4.17: A trihedron move

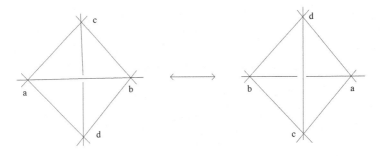

Figure 4.18: A tetrahedron move

Definition 4.24 *A* trihedron move *is shown in Fig. 4.17.*

Definition 4.25 *A* tetrahedron move *is shown in Fig. 4.18.*

The rest of $TL(rot(\hat\beta)) \hookrightarrow S^1 \times S^1 \times \mathbb{R}^+$ is unchanged under the moves. The stickers on the edges change in the canonical way.

Notice, that a trihedron move corresponds to a generic homotopy of the canonical loop which passes once through an ordinary tangency with a stratum of $\sum_{tri}^{(1)}$. A tetrahedron move corresponds to a generic homotopy of the canonical loop which passes transversely once through a stratum of $\sum_{quad}^{(2)}$, i.e. corresponding to an ordinary quadruple point in the projection.

Definition 4.26 *The* equivalence relation *for trace graphs* $TL(rot(\hat\beta))$ *is generated by the following three operations:*
(1) isotopy in the thickened torus
(2) trihedron moves
(3) tetrahedron moves.

The following important Reidemeister type theorem is a particular case of Theorem 1.10 in [20], compare also Theorem 4.1 in Section 4.1.3.

Theorem 4.4 *(Higher order Reidemeister theorem for trace graphs of closed braids)*

Two closed braids (which are knots) are isotopic in the solid torus if and only if their trace graphs in the thickened torus are equivalent.

Remark 4.1 *Notice, that not all representatives of an equivalence class of a trace graph correspond to the canonical loop (which is very rigid, because it is determined by a single braid diagram in it) of some closed braid. However, one easily sees that all representatives correspond to loops in the space $M(\hat{\beta})$.*

Remark 4.2 *A generic homotopy of loops $\gamma_s, s \in [0,1]$, in $M(\hat{\beta})$ could of course become tangential for some s to $\sum_{tan}^{(1)}$ at a generic point. We can allow tangencies of $\sum_{tan}^{(1)}$ on the negative side, i.e. where the ordinary diagrams have two crossings less, because this corresponds to the birth or the death of a (null-homologous) component of the trace graph. However, a tangency on the positive side would imply a Morse modification of index 1 of the trace graph and, hence, change the components of its natural resolution in an uncontrollable way. The important point is, that this does not happen for the (very rigid) homotopies of $rot(\hat{\beta})$ which are induced by generic isotopies of $\hat{\beta}$ in V, compare Observation 1.1.*

Proof of Observation 1.1 from Chapter 1. Indeed, under a monotonous rotation of the closed braid around the core of the solid torus V, $rot(\hat{\beta})$ is tangential to $\sum_{tan}^{(1)}$ in an ordinary point if and only if the trace graph $TL(rot(\hat{\beta}))$ has a Morse singularity. This could only happen if two strings of the braid would be become tangential to the same disc $(\phi = const) \times D^2$ in the fibration of V. But this is not possible, because the closed braids are always transverse to the disc-fibration of V. □

Consequently, the trace circles of (the resolution of) the trace graph are isotopy invariants for closed braids.

Remark 4.3 *There can be loops γ_s in a homotopy which are tangential to $\sum_{tri}^{(1)}$. For the trace graphs associated to the loops γ_s this corresponds to a trihedron move (compare [20] and [21]).*

We have some more information about isotopies of trace graphs.

Figure 4.19: A triple point slides over a vertical tangency

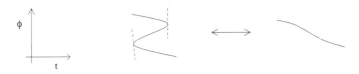

Figure 4.20: A couple of tangencies appears or disappears

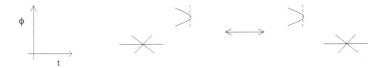

Figure 4.21: A commutation of a triple point with a vertical tangency

Definition 4.27 *A time section in the thickened torus* $(S^1_\phi \times \mathbb{R}^+_\rho) \times S^1_t$ *is an annulus of the form* $(S^1_\phi \times \mathbb{R}^+_\rho) \times \{t = const\}$.

The intersection of $TL(rot(\hat{\beta}))$ with a generic time section corresponds to the crossings of the closed braid $\hat{\beta}_t$. Using the orientation and the stickers on $TL(rot(\hat{\beta}))$ we can read off a cyclic braid word for $\hat{\beta}$ in each generic time section. The tangent points of $TL(rot(\hat{\beta}))$ with time sections correspond exactly to the Reidemeister II moves in the one parameter family of diagrams $\hat{\beta}_t, t \in S^1$. A triple point in $TL(rot(\hat{\beta}))$ slides over such a tangent point if and only if the canonical loop passes in a homotopy transversely through a stratum of $\sum^{(2)}_{trans-self}$. We illustrate this in Fig. 4.19.

When the canonical loop passes transversely through a stratum of $\sum^{(2)}_{self-flex}$ then the trace graph changes as shown in Fig. 4.20.

Finally, when the canonical loop passes transversely through a stratum of $\sum^{(2)}_{inter}$ then the t-values of triple points or tangencies with time sections are interchanged. We show an example in Fig. 4.21.

It turns out that we could replace Observation 1.1 by the following observation of Orevkov, but at the cost that braids become much longer.

Remember that in our one-parameter theory we consider (geometric) braids as tangles and not as isotopy classes of tangles (or elements in the braid group). So, a generic braid is for us a word in the standard generators and their inverses of the braid group.

If a braid β is in *Garsides normal form*, see [23], then it is in particular of the form $\beta^+(\Delta^2)^k$, where β^+ is a positive braid, k is an integer and β^+ is not divisible by Δ^2.

Let $\beta^+(\Delta^2)^k$ and $\beta'^+(\Delta^2)^k$ be two n-braids in Garsides normal form. *Garsides solution of the conjugacy problem* says that the the isotopy classes of the braids are conjugate in B_n if and only if β^+ and β'^+ can be related by a sequence of the following operations (see [23]):

1) isotopy amongst positive braids

2) cyclic permutation

3) conjugacy with permutation braids $s\beta^+s^{-1}$ followed by a simplification to a positive braid (where the *permutation braids* s are exactly the left and right divisors in the braid monoid B_n^+ of Garsides braid Δ, i.e. there exist positive n-braids s' and s'' such that ss' is isotopic to $s''s$ which is isotopic to Δ).

Stepan Orevkov has made the following observation.

Observation 4.1 *(Orevkov) Assume that the closure $\hat{\beta}^+$ contains a half-twist Δ, i.e. β^+ is equivalent to $\beta''^+\Delta$ for some positive braid β''^+ by operations 1 and 2 only. Then we can skip the operation 3 in Garsides solution of the conjugacy problem.*

Proof. Indeed, assume that we have to perform the operation 3:

$$\beta^+ \to s\beta^+s^{-1} = \beta'^+.$$

From our hypothesis $\beta^+ = \beta''^+\Delta$ by using operations 1 and 2. $\beta''^+\Delta = \beta''^+s''s \to s\beta''^+s''ss^{-1} = s\beta''^+s'' = \beta'^+$. But by using only operations 2 we can bring the last braid to $\beta''^+s''s = \beta''^+\Delta = \beta^+$. Consequently, we have replaced the operation 3 by a sequence of only operations of type 1 and 2. $\square$

Orevkov's observation implies that if two positive closed braids which contain a half-twist are isotopic in the solid torus then there is such an isotopy which stays within positive closed braids (because negative crossings occur only in operation 3). The loop $rot(\hat{\beta}''^{+}\Delta)$ can be represented by pushing twice Δ through β''^{+}. It follows easily now (by considering braids as diffeomorphisms of the punctured disc and *rot* as the rotation from 0 to 2π of the disc around its center) that if $\hat{\beta}''^{+}\Delta$ is isotopic to $\hat{\beta}'^{+}\Delta$ then the loops $rot(\hat{\beta}''^{+}\Delta)$ and $rot(\hat{\beta}'^{+}\Delta)$ are homotopic through positive closed braids (i.e. all generic crossings are positive). Consequently, there are no Morse modifications of the trace graph at all in such a homotopy and hence the trace circles are again isotopy invariants of the closed braid.

Adding full-twists to a braid does not change its geometry and its geometric invariants (as hyperbolic volume or entropy for pseudo-Anosov braids) and a full twist (which generates the center of the braid group) in a closed braid can be easily detected. Using Orevkov's observation we could therefore restrain our-self to the heart of the matter: constructing 1-cocycles for positive closed braids in $M_{\hat{\beta}+\Delta}$, but of course braids would be in general much longer. Therefore we allow in this book negative crossings as well.

4.8 Polynomial one-cocycles for closed braids in the solid torus

In this section we introduce our one-cocycle polynomials for closed braids. Notice, that we do not need trace graphs here. They will only be needed in the refinements of finite type invariants to character invariants in the next section.

Our strategy is the following: for an oriented generic loop or arc in $M_n^{+} \subset M_n$ (compare the Introduction) we associate some polynomial to the intersection with each stratum in $\Sigma_{tri}^{(1)}$, i.e. to each Reidemeister move of type III, and we sum up over all Reidemeister moves in the arc. It could be seen as an integration of a discrete closed 1-form over a discrete loop. We have to prove now that this sum is 0 for each meridian of strata in $\Sigma^{(2)}$, compare Proposition 1.1. It follows that our sum is invariant under generic homotopies of arcs (with fixed endpoints). But it takes its values in an abelian ring and hence it is a 1-cocycle.

4.8.1   Gauss diagrams for closed braids with a triple crossing

Let $f : S^1 \to \hat{\beta}$ be a generic orientation preserving diffeomorphism. Let p be any crossing of $\hat{\beta}$. We connect $f^{(-1)}(p) \in S^1$ by an oriented arrow, which goes from the under-cross to the over-cross and we decorate it by the writhe $w(p)$. Moreover, we attache to the oriented chord the homological marking $[p^+]$ (compare Section 4.7.2). The result is called a *Gauss diagram* for $\hat{\beta}$ (compare also e.g. [47], [16] and Section 2.1).

One easily sees that $\hat{\beta}$ up to isotopy is determined by its Gauss diagram and the number $n = [\hat{\beta}] \in H_1(V)$.

We can form oriented loops in Gauss diagrams of closed braids, which are knots, in the following way: going along the circle following its orientation we can possibly jump at arrows and continue going along the circle following its orientation up to reaching our starting point. The following simple observation is at the origin of our new solution of the tetrahedron equation which leads immediately to our polynomial valued one-cocycles.

Observation 4.2 *Each such loop in the Gauss diagram of a closed braid, which is a knot, represents a homology class in* $\{1, 2, \ldots, n - 1\}$.

Proof. Each such loop is positive transverse to the disc fibration of V exactly as the closed braid, besides at the jumps, which can be represented by arcs in the discs. Consequently, the loop represents a homology class between 1 and $n - 1$. $\square$

In other words, $\hat{\beta} \in M_n^+$ instead of just M_n.

A *Gauss sum of degree d* is an expression assigned to a diagram of a closed braid which is of the following form, compare Section 2.1 and [16]:

$$\sum \text{function(writhes of the crossings)}$$

where the sum is taken over all possible choices of d (unordered) different crossings in the knot diagram such that the arrows without the writhes arising from these crossings build a given sub-diagram with given homological markings. The marked sub-diagrams (without the writhes) are called *configurations*. If the function is the product of the writhes (this is always the case in this paper), then we will denote the sum shortly by the configuration itself. We need to define Gauss diagrams for knots with an ordinary triple

Figure 4.22: The coorientation for a triple crossing for closed braids

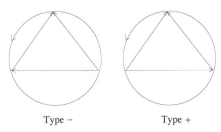

Type −   Type +

Figure 4.23: The unmarked global types of triple crossings

point too. The triple point corresponds to a triangle in the Gauss diagram of the knot. Notice, that the preimage of a triple point has a natural ordering coming from the orientation of the $\mathbb{R}^+$-factor. One easily sees that this order is completely determined by the arrows in the triangle.

We provide each stratum of $\sum_{tri}^{(1)}$ with a co-orientation which depends only on the non-oriented underlying curves $pr(\hat{\beta})$ in $S^1 \times \mathbb{R}^+$. Consequently, for the definition of the co-orientation we can replace the arrows in the Gauss diagram simply by chords.

Definition 4.28 *The* co-orientation *of the strata in* $\sum_{tri}^{(1)}$ *is given in Fig. 4.22.*

Let s be a point in the transverse intersection of an oriented arc $S \subset M_n$ with $\sum_{tri}^{(1)}$. Then the sign $sign(s)$ is $+1$ if the orientation of S agrees with the co-orientation of $\sum_{tri}^{(1)}$ at s and $sign(s) = -1$ otherwise.

There are exactly two types of triple points without markings. We show them in Fig. 4.23, where l is replaced by $-$ and r is replaced by $+$.

We attach now the homological markings to the three chords. Let $a, b \in \{1, 2, \ldots, n-1\}$ be fixed. Then the markings of a triple crossing are as shown in Fig. 4.24. We encode the types of the marked triple points by $(a, b)^-$ and respectively $(a, b)^+$. The union of the corresponding strata of $\sum_{tri}^{(1)}$ are encoded in the same way.

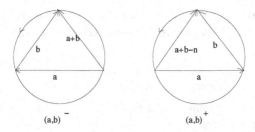

$$(a,b)^- \qquad\qquad (a,b)^+$$

Figure 4.24: The marked global types of triple crossings for closed braids

As for knots in the solid torus (compare Chapter 2, Fig. 2.2) we denote by d the crossing between the highest and the lowest branch in the triple crossing, by hm the crossing between the highest and the middle branch and by ml the crossing of the middle branch with the lowest branch.

4.8.2 The general form of one-cocycle polynomials from Gauss diagrams with a triple crossing

We will construct one-cocycles on the space M_n^+ (the space of all closed n-braids which are knots). We obtain invariants of closed braids when we evaluate these cocycles on the homology class represented by the canonical loop *rot*. (If a braid β is reducible then there is an incompressible not-boundary parallel torus in the complement $V \setminus \hat{\beta}$. This implies that there are new loops in $M_{\hat{\beta}}^+$ to which we could apply our one-cocycles as well, compare [29] for the case of long knots.)

Let $n, d \in \mathbb{N}^*$ be fixed. Let $(a,b)^+$, respectively $(a,b)^-$, be a fixed *type of marked triple point* as shown in Fig. 4.24.

Definition 4.29 *A configuration I of degree d is an abstract Gauss diagram without writhes which contains exactly d arrows marked in $\{1, 2, \ldots, n-1\}$ besides the triangle $(a,b)^+$ or $(a,b)^-$.*

Let $\{I_i\}$ be the finite set of all configurations of degree d with respect to $(a,b)^{\pm}$. Let $\Gamma = \sum_i \epsilon_i I_i$ be a given linear combination with each $\epsilon_i \in \{0, +1, -1\}$. (The type of the triple point is *always* fixed in any cochain Γ).

Here is the general form of our cochains.

Definition 4.30 Γ gives rise to a 1-cochain by assigning to each oriented generic loop $S \subset M_n^+$ an integer Laurent polynomial $\Gamma(S)$ in the following way:

$$\Gamma(S) = \sum_{\substack{s_i \in S \cap \Sigma_{tri}^{(1)} \, of \, type \, (a,b)^\pm}} sign(s_i) x^{\sum_i \epsilon_i (\sum_{D_i} \prod_j w(p_j))}$$

where D_i is the set of unordered d-tuples $(p_1, \ldots, p_d)$ of arrows which enter in I_i in the Gauss diagram of s_i.

Lemma 4.21 If $\Gamma(S)$ is invariant under each generic deformation of S through any stratum of $\Sigma^{(2)}$, then Γ is a 1-cocycle.

Proof. In this case, $\Gamma(S)$ is invariant under homotopies of S. Indeed, tangent points of S with $\Sigma_{tri}^{(1)}$ correspond just to trihedron moves. The two triple points give the same contribution to $\Gamma(S)$ but with different signs. A tangency with $\Sigma_{tan}^{(1)}$ does not change the contribution of the triple points at all. This implies the invariance under homologies of S because $\Gamma(S)$ takes it values in an abelian ring. $\square$

Definition 4.31 A cohomology class in $H^1(M_n^+; \mathbb{Z}[x, x^{-1}])$ is of Gauss degree d if it can be represented by some 1-cocycle Γ such that each configuration I_i has at most d arrows besides the three arrows of the triangle.

Remark 4.4 The integer valued invariant $d(\Gamma(rot(\hat{\beta}))/dx$ at $x = 1$ is given by a Gauss diagram formula and it is hence an invariant of finite type. Indeed, the one-cocycle invariant $\Gamma(rot(\hat{\beta}))$ is calculated as some sum $\sum_{s_i}$ over triple points s_i in $rot(\hat{\beta})$. Therefore, it suffices to prove that this sum $\sum_{s_i}$ for each triple point s_i is of finite type (even if it is not invariant). If Γ is of Gauss degree d then $\sum_{s_i}$ depends only on the triple point and of configurations of d other crossings. This means that in order to calculate a summand in $\sum_{s_i}$ we can switch all other crossings besides the triple point and the fixed d crossings. The result will not change. This implies immediately that each $\sum_{s_i}$ is of degree at most $d + 1$ (see [47] and also [16]).

The above definition induces a filtration on a part of $H^0(M_n^+; \mathbb{Z})$ by taking all $d(\Gamma(rot(\hat{\beta})))/dx|_{x=1}$.

Let X be the (disconnected) space of all embeddings $f : S^1 \hookrightarrow \mathbb{R}^3$. Vassiliev [54] has introduced a filtration on a part of $H^0(X; \mathbb{Z})$ using the discriminant Σ_{sing} of singular maps. It is not difficult to see that each component of the space of all (unparametrized) differentiable maps of the circle into the solid torus is contractible. Indeed, there is an obvious canonical homotopy of each (perhaps singular) knot to a multiple of the core of the solid torus. The core of the solid torus is invariant under $rot_{S^1}(V) \times rot_{D^2}(V)$. Thus, the above space is star-like. Therefore, Alexander duality could be applied and Vassilievs original approach could be generalized for knots in the solid torus too. It would be interesting to compare his filtration with our filtration (we compare them just for the Gauss degree 0 in Section 4.8.3). In the next sections, we will construct 1-cocycle polynomials Γ in an explicit way. We show with a simple example that they are in general not of finite type.

4.8.3   One-cocycles of Gauss degree 0

Let $\beta \in B_n$ be such that its closure $\hat{\beta} \hookrightarrow V$ is a knot.

Let us recall the simplest finite type invariants for closed braids in Vassiliev's sense.

Proposition 4.7 *The space of finite type invariants of degree 1 is of dimension $[n/2]$ (here $[.]$ is the integer part). It is generated by the Gauss diagram invariants $W_a(\hat{\beta}) = \sum w(p)$, where $a \in \{1, 2, \ldots, [n/2]\}$. The sum is over all crossings with fixed homological marking a.*

Proof. It follows from Goryunov's [25] generalization of finite type invariants for knots in the solid torus (likewise by generalizing the Kontsevich integral or by generalizing the defining skein relations for finite type invariants) that the invariants of degree 1 correspond just to marked chord diagrams with only one chord. Obviously, all these invariants can be expressed as Gauss diagram invariants:

$$W_a(\hat{\beta}) = \sum w(p), \ a \in \{1, \ldots, n-1\}$$

(see Section 2.2 in [16]).

Let us define $V_a(\hat{\beta}) := W_a(\hat{\beta}) - W_{n-a}(\hat{\beta})$ for all $a \in \{1, \ldots, n-1\}$. We observe that $V_a(\hat{\beta})$ is invariant under switching crossings of $\hat{\beta}$. Indeed, if the marking of the crossing p was $[p] = a$, then the switched crossing p^{-1} has marking $[p^{-1}] = n - a$, but $w(p) = -w(p^{-1})$. But every braid $\beta \in B_n$ is homotopic to $\gamma = \prod_{i=1}^{n-1} \sigma_i$. A direct calculation for γ shows that $V_a(\hat{\gamma}) \equiv 0$. It is easily shown by examples that $W_a, a \in \{1, \ldots, [n/2]\}$ (seen as invariants in with values in $\mathbb{Q}$) are linearly independent. $\square$

Figure 4.25: Different triple crossings come together in an self-tangency

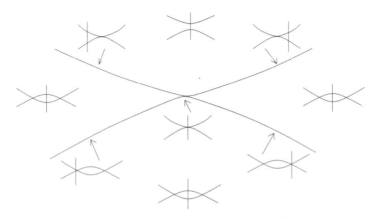

Figure 4.26: The unfolding of an self-tangency with a transverse branch

Lemma 4.22 *Let $a, b \in \{1, 2, \ldots, n-1\}$ be fixed. Consider the union of all co-oriented strata of $\Sigma^{(1)}$ which correspond to triple points of type either $(a, b)^-$ or $(a, b)^+$. The closure in M_n^+ of each of these sets form integer cycles of codimension 1 in M_n.*

Remark 4.5 *Otherwise stated, $\Gamma_{(a,b)^+}$ and $\Gamma_{(a,b)^-}$ both define integer valued 1-cocycles of Gauss degree 0. $\Gamma_{(a,b)^\pm}(S)$ is in this case by definition just the algebraic intersection number of S with the corresponding union of strata of $\Sigma_{tri}^{(1)}$. The variable x enters the one-cocycles only starting from the Gauss degree 1.*

Proof. According to Section 4.2, we have to prove that the co-oriented strata fit together in $\Sigma_{quad}^{(2)}$, $\Sigma_{trans-self}^{(2)}$ and $\Sigma_{inter}^{(2)}$. The first is evident, because at a stratum of $\Sigma_{quad}^{(2)}$ just four strata of $\Sigma_{tri}^{(1)}$ intersect pairwise transversely. For the second, we have to distinguish 24 cases. Three of them are illustrated in Fig. 4.25.

The whole picture in a normal 2-disc of $\Sigma^{(2)}_{trans-self} \subset M_n^+$ is then shown in Fig. 4.26, but we draw only the corresponding planar curves.

All other cases are obtained from these three by inverting the orientation of the vertical branch, by taking the mirror image (i.e. switching all crossings), and by choosing one of two possible closings of the 3-tangle (in order to obtain an oriented knot). In all cases, one easily sees that the two adjacent triple points are always of the same marked type and that the co-orientations fit together. □

Proposition 4.8 $\Gamma_{(a,b)+}$ *defines a non-trivial 1-cohomology class of Gauss degree 0 if and only if $a \neq b$ and $a + b \leq n - 1$.*

$\Gamma_{(a,b)-}$ *defines a non-trivial 1-cohomology class of Gauss degree 0 if and only if $a \neq b$ and $a + b \geq n + 1$.*

The following identities hold:

(*) $$\Gamma_{(a,b)+} + \Gamma_{(b,a)+} \equiv 0$$
(*) $$\Gamma_{(a,b)-} + \Gamma_{(b,a)-} \equiv 0$$

Proof. For closed braids, the markings are all in $\{1, \ldots, n-1\}$. Therefore, if $a + b > n - 1$ in $\Gamma_{(a,b)+}$ or $a + b < n + 1$ in $\Gamma_{(a,b)-}$, then there is no such triple point at all and the 1-cocycle is trivial. It follows from the identities that $\Gamma_{(a,a)+}$ and $\Gamma_{(a,a)-}$ are trivial.

Examples show that all the remaining 1-cocycles are non-trivial.

In order to prove the identities, we use the following Gauss diagram sums (see also Section 1.6 in [16]):

$$I_{(a,b)}^+ = \sum w(p)w(q), \qquad I_{(a,b)}^- = \sum w(p)w(q).$$

Here, the first sum is over all couples of crossings which form a subconfiguration as shown on the left in Fig. 4.27. The second sum is over all couples of crossings which form a sub-configuration as shown on the right in Fig. 4.27. (Here, a and b are the homological markings.)

These sums applied to diagrams of $\hat{\beta}$ are not invariants. Let $S \subset M_{\hat{\beta}}^+$ be a generic loop. Then $I_{(a,b)}^\pm$ is constant except when S crosses $\Sigma_{tri}^{(1)}$ in strata of type $(a, b)^\pm$ or $(b, a)^\pm$. At each such intersection in positive (respectively negative) direction, $I_{(a,b)}^\pm$ changes exactly by -1 (respectively $+1$). Indeed, the configurations of the three crossings which come together in a triple point are shown in Fig. 4.28.

Figure 4.27: Two sub-diagrams

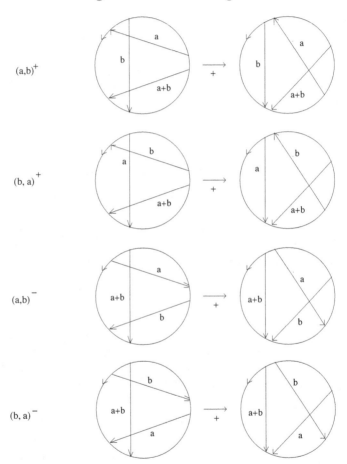

Figure 4.28: Perturbations of triple crossings

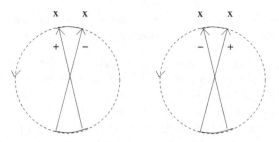

Figure 4.29: Contributions of self-tangencies

In each of the four cases, exactly one pair p, q of crossings contributes to one of the sums I.

After drawing all possible triple points, it is easily seen that p, q must verify: $w(p)w(q) = -1$ for the first two cases and $w(p)w(q) = +1$ for the last two cases. Notice that the type of the triple point is completely determined by the sub-configurations shown in Fig. 4.27. Thus, the sums I are constant by passing all types of triple points except those shown in Fig. 4.28. The generic loop S intersects Σ only in strata that correspond to triple points or to self-tangencies. A self-tangency adds to the Gauss diagram always one of the sub-diagrams shown in Fig. 4.29.

The two arrows evidently do not enter together in the configurations shown in Fig. 4.27. If one of them contributes to such a configuration, then the other contributes to the same configuration but with an opposite sign.

Therefore, for any two $\hat{\beta}_1, \hat{\beta}_2 \in M_{\hat{\beta}}^+$, the difference $I_{(a,b)}^{\pm}(\hat{\beta}_1) - I_{(a,b)}^{\pm}(\hat{\beta}_2)$ is just the algebraic intersection number of an oriented arc from $\hat{\beta}_1$ to $\hat{\beta}_2$ with the union of the cycles of codimension one $(a, b)^+ \cup (b, a)^+$ (resp., $(a, b)^- \cup (b, a)^-$). Hence, for each loop S, these numbers are 0. The identities (*) follow. $\square$

Example 4.2 *Let $\hat{\beta}$ be the closure of the 4-braid $\beta = \sigma_1 \sigma_2^{-1} \sigma_3^{-1}$. We consider $\Gamma_{(1,2)^-}, \Gamma_{(2,1)^-}, \Gamma_{(2,3)^+}$ and $\Gamma_{(3,2)^+}$.*

A calculation by hand gives:

$$\Gamma_{(1,2)^-}(rot(\hat{\beta})) = \Gamma_{(2,3)^+}(rot(\hat{\beta})) = -1$$

and

$$\Gamma_{(2,1)^-}(rot(\hat{\beta})) = \Gamma_{(3,2)^+}(rot(\hat{\beta})) = +1.$$

Therefore, all four 1-cocycles of Gauss degree 0 are non-trivial and they are obviously related to the finite type invariants $W_a(\hat{\beta})$ of degree 1.

The non-triviality of a one-cocycle invariant on the loop $rot(\hat{\beta})$ implies that the braid β is not periodic, because of the following result of Hatcher.

Let $slide(\hat{\beta})$ be the loop which is obtained by the rotation of the solid torus around the core of the complementary solid torus $S^3 \setminus V$. Evidently, $slide(\hat{\beta})$ is just a rotation of diagrams in the annulus and hence it does not intersect the discriminant $\sum^{(i)}, i > 0$, at all. Consequently, each one-cocycle invariant vanishes on this loop. The following proposition was proven by Allen Hatcher.

Proposition 4.9 *(Hatcher). The loops $rot(\hat{\beta})$ and $slide(\hat{\beta})$ represent linearly dependent homology classes in $H_1(M_{\hat{\beta}}; \mathbb{Q})$ if and only if $\hat{\beta}$ is isotopic to a torus knot in ∂V.*

Proof. The rotations give an action of $\mathbb{Z} \times \mathbb{Z}$, and one can look at the induced action on the fundamental group of the knot complement, with respect to a base point in the boundary torus ∂V of the solid torus. Call this fundamental group G. The action of $\mathbb{Z} \times \mathbb{Z}$ on G is conjugation by elements of the $\mathbb{Z} \times \mathbb{Z}$ subgroup of G represented by loops in the boundary torus. The action will be faithful (zero kernel) if the center of G is trivial. The only (irreducible) 3-manifolds whose fundamental groups have a nontrivial center are Seifert fibered manifolds defined by a circle action. The only such manifolds that embed in $\mathbb{R}^3$ and have two boundary components (as here) are the obvious ones that correspond to cable knots, that is, the complement of a knot in a solid torus that is isotopic to a nontrivial loop in the boundary torus. □

On the other hand, it is well known that a braid β, which closes to a knot, is *periodic* if and only if $\hat{\beta}$ is isotopic to a torus knot in ∂V (see e.g. [7]). Consequently, if some one-cocycle evaluated on $rot(\hat{\beta})$ is non-trivial, then the braid β is not periodic. The above braids are evidently not reducible and as well known this implies now that they are pseudo-Anosov, compare [51], [6].

It would be very interesting to find out which information about the entropy of a braid β and about the simplicial volume of its mapping torus $S^3 \setminus (\hat{\beta} \cup L)$ can be obtained from the set of all one-cocycle polynomials $\Gamma(rot(\hat{\beta}))$ (compare the next section). But unfortunately a computer program is still missing in order to calculate lots of examples and to make precise conjectures.

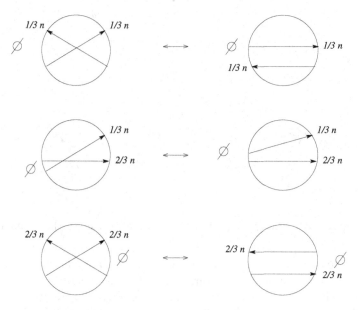

Figure 4.30: Slide moves for a couple of arrows not in the triangle

4.8.4   One-cocycles of higher Gauss degree

A general procedure in order to define all one-cocycle invariants with our method was given in the preprint [17]. However, it seems to be not very understandable. Therefore we restrict our-self in this section to the construction of very simple examples, which illustrate our method and which can be generalized easily by an interested reader.

Let $\beta \in B_n$ such that $\hat{\beta} \hookrightarrow V$ is a knot. We assume that n is divisible by 3. We consider only the type of the triple crossings $(2/3n, 2/3n)^+$ and all markings in the configurations have to be of type $1/3n$ or $2/3n$.

We define a *slide move for a couple of arrows* in a configuration in Fig. 4.30. Here $\emptyset$ means that there are no heads or foots of arrows of the configuration in the corresponding segment on the circle.

We define *slide moves for an arrow with respect to the triangle* in Fig. 4.31. Here the condition is that the move does not create an oriented loop in the diagram which represents a homology class which is not in $\{1, 2, \ldots, n-1\}$, compare Observation 4.2. (The marking a here is $1/3n$ or respectively $2/3n$.)

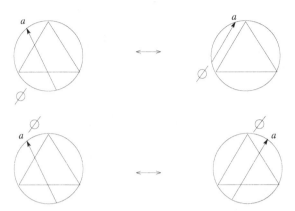

Figure 4.31: Slide moves of an arrow with respect to the triangle

We define an *exchange move for an arrow with the triangle* in Fig. 4.32. Notice that the sign of the configuration changes in this case.

We define a *forbidden sub-configuration* in Fig. 4.33, where the arrows have the same homological marking a. (The marking a here is $1/3n$ or respectively $2/3n$.)

Theorem 4.5 *Let $I^d = \sum_i \epsilon_i I_i$ be a linear combination of configurations of Gauss degree d and which is invariant under all possible slide moves, exchange moves and which does not contain any forbidden sub-configuration. Then*

$$\Gamma^d(S) = \sum_{s_i \in S \cap \Sigma^{(1)}_{tri} \, of \, type \, (2/3n, \, 2/3n)^+} sign(s_i) x^{\sum_i \epsilon_i (\sum_{D_i} \prod_j w(p_j))}$$

is a one-cocycle, where D_i is the set of unordered d-tuples $(p_1, \ldots, p_d)$ of arrows which enter in I_i in the Gauss diagram of s_i.

Proof. Following Proposition 1.1 and Lemma 4.21, we have to show that Γ^d vanishes on all meridians of $\sum^{(2)} = \sum^{(2)}_{quad} \cup \sum^{(2)}_{trans-self} \cup \sum^{(2)}_{self-flex} \cup \sum^{(2)}_{inter}$ in M_n.

The Gauss diagrams in the meridian of $\sum^{(2)}_{inter}$ for two triple crossings change just by slide moves for a couple of arrows and the two triple crossings enter with different signs, compare [20]. Consequently $\Gamma^d = 0$ on the meridian. In the transverse intersection of $\sum^{(1)}_{tri}$ with $\sum^{(1)}_{tan}$ there appear two

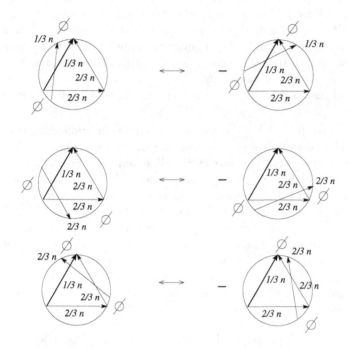

Figure 4.32: Exchange move with an arrow in the triangle

Figure 4.33: A forbidden sub-configuration

new arrows in the Gauss diagram of one of the two triple crossings in the meridian. These two arrows do not enter together into any configuration because I^d does not contain any forbidden sub-configurations. The two arrows have different signs and consequently, if one of the arrows enter into a configuration then the other enters too but with a different sign of the weight and their contributions cancel out.

Self-tangencies do not contribute to Γ^d and hence $\Gamma^d = 0$ on the meridian of $\sum_{self-flex}^{(2)}$.

In the meridian of $\sum_{quad}^{(2)}$ there are exactly eight triple crossings. They come in pairs with different signs, compare [20] and also Fig. 1.2. The Gauss diagrams of the pairs differ just by the slide moves of three different arrows with respect to the triangle and hence again $\Gamma^d = 0$ on the meridian, i.e. Γ^d is a solution of the tetrahedron equation. In the meridian of $\sum_{trans-self}^{(2)}$ there are exactly two triple crossings and they have different signs. The two Gauss diagrams of the braids with the triple crossing differ exactly by an exchange move of one arrow with the triangle. Moreover, the sign of the arrow changes (but we do not care about the signs of the arrows in the triangle), compare [20]. But I^d stays invariant, because the sign of the configuration changes too for the exchange move with a triangle. Consequently, $\Gamma^d = 0$ on the meridian. □

We give three examples in Fig. 4.34, Fig. 4.35 and Fig. 4.36.

Let us give names to the three strands in a triple crossing: h is the highest strand, m is the strand in the middle and l is the lowest strand. In each of the examples, the bunch of alternating arrows separates one of the strands from the other two. We use the single separated strand in the notation of the configuration (compare the figures).

The three configurations in the examples define one-cocycles because of the following proposition.

Proposition 4.10 *The one-cochains I_h^d for even d and I_l^d for all d are invariant under all possible slide moves, exchange moves and do not contain any forbidden sub-configurations.*

Proof. I_h^d and I_l^d for even d are *rigid*, i.e. no slide move or exchange move at all is possible. Indeed, any slide move would create a loop in the Gauss diagram with homological marking 0. But this contradicts Observation 4.2. There is no arrow at all which would allow an exchange move with the triangle

$$I^{d}_{(2/3\,n,\,2/3\,n)^{+},\,l} =$$

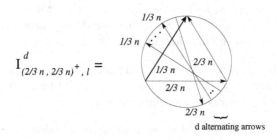

Figure 4.34: Configuration for even d

$$I^{d}_{(2/3\,n,\,2/3\,n)^{+},\,h} =$$

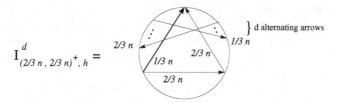

Figure 4.35: Another configuration for even d

$$I^{d}_{(2/3\,n,\,2/3\,n)^{+},\,l} =$$

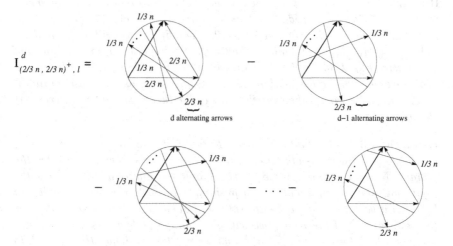

Figure 4.36: Configuration for odd d

because of the markings. For the same reason, I_l^d for odd d allows only slide moves of just one arrow, and one easily sees that I_l^d takes into account all possible slide moves of this arrow, called *wandering* arrow, as well as the single possible exchange move for the wandering arrow. □

But remember that the alternating arrows form only a configuration, i.e. in the Gauss diagram of the closed braid there could be other arrows which cut the alternating arrows in an arbitrary way.

Remark 4.6 *The one-cochain which is obtained by taking mirror images of everything in I^d, i.e. changing the orientation of each arrow and interchanging the markings $1/3n$ with $2/3n$, is of course also a one-cocycle.*

If n is not divisible by 3 then we replace $\hat{\beta}$ by a 3k-cable, $k \in \mathbb{N}^*$, which is twisted by the permutation braid $\sigma_1\sigma_2 \ldots \sigma_{3k-1}$ in order to get a knot, and we can take now the markings kn and $2kn$. Indeed, we can imagine the untwisted 3k-cable as 3k parallel strands on a band (which projects into the annulus by an immersion) and we can push the permutation braid along the band. Hence, if two closed braids are isotopic in the solid torus then their 3k-cables which are twisted by the same braid, are still isotopic.

Remark 4.7 *Let 1-cocycles Γ^d, $d > 0$, depend on the order of the Reidemeister moves in the loop. Indeed, let us consider a move $(1,1)^-$ followed by a move $(2,2)^+$ for $n = 3$, compare Fig. 4.24. The first move will not contribute to $\Gamma^1_{(2,2)^+,l}(rot(\hat{\beta}))$, but it changes the weight for the crossings ml of marking 1, compare Fig. 4.36 for $d = 1$. Consequently, if we change the order of the two moves and the crossing d in $(2,2)^+$ is the same as the crossing ml in $(1,1)^-$, then the sum of the contributions of the two moves to $\Gamma^1_{(2,2)^+,l}(rot(\hat{\beta}))$ changes.*

Example 4.3 *Notice that for closed 3-braids, the homological markings of the triple point are already determined by the arrows. Let us consider the 3-braid $\beta = \sigma_1\sigma_2^{-1}\sigma_1\sigma_2\sigma_1\sigma_1\sigma_2\sigma_1$. The braid β contains a full-twist at the right end, and we represent rot by pushing it through $\sigma_1\sigma_2^{-1}$ to the left and bringing it back to the right by an isotopy of diagrams in the solid torus. For the convenience of the reader we will give the loop below. As usual we will write shortly i for σ_i and $\bar{i}$ for σ_i^{-1} and we put the Reidemeister moves into brackets.*

$1\bar{2}(121)121 \to 1(\bar{2}\bar{2})12121 \to 1(2\bar{2})12121 \to 12(\bar{2}12)121 \to 1212(\bar{1}1)21 \to$
$1212(1\bar{1})21 \to 12121(\bar{1}21) \to 121(212)1\bar{2} \to 12112\bar{1}1\bar{2} \to 1\bar{2}121121$

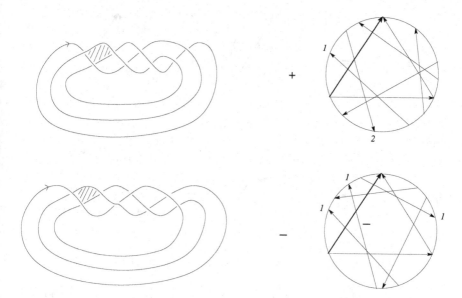

Figure 4.37: Contributing triple crossings with their Gauss diagrams

Here, the last arrow is an isotopy of diagrams. It turns out that only the second and the forth Reidemeister III move is of type $(2,2)^+, l$. We show the corresponding closed braids together with their Gauss diagrams in Fig. 4.37.

We indicate the sign of the move, the homological markings of the relevant crossings and the sign of the crossings, but only if it is negative. One easily calculates now that

$$\Gamma^2_{(2,2)+,l}(rot(\hat{\beta})) = x \text{ and } \Gamma^1_{(2,2)+,l}(rot(\hat{\beta})) = x - x^{-1}.$$

Example 4.4 *Let us consider the closure of the 3-braid $\beta_{+++} = 11122121$. There are exactly eight Reidemeister III moves in pushing 121 twice through the rest of the closed braid. A calculation gives now*

$$\Gamma^4_{(2,2)+,h}(rot(\hat{\beta}_{+++})) = -x, \quad \Gamma^2_{(2,2)+,h}(rot(\hat{\beta}_{+++})) = -x^3 - x,$$

$$\Gamma^4_{(2,2)+,l}(rot(\hat{\beta}_{+++})) = x, \quad \Gamma^2_{(2,2)+,l}(rot(\hat{\beta}_{+++})) = x^3 + x,$$

$$\Gamma^1_{(2,2)+,l}(rot(\hat{\beta}_{+++})) = x^2 + x - x^{-1} - x^{-2}.$$

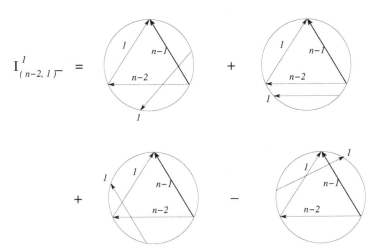

Figure 4.38: A configuration of Gauss degree 1 for $n > 3$

Let us consider $\beta_{-++} = 11\bar{1}22121$, $\beta_{+-+} = 111\bar{2}2121$, $\beta_{--+} = 11\bar{1}\bar{2}2121$, $\beta_{++-} = 1112212\bar{1}$, $\beta_{-+-} = 11\bar{1}2212\bar{1}$, $\beta_{+--} = 111\bar{2}212\bar{1}$, $\beta_{---} = 11\bar{1}\bar{2}212\bar{1}$.

If an invariant $\Gamma(rot(\hat{\beta}))$ would be of finite type of degree 2 then we would have (compare [2], [54])

(*) $\Gamma(rot(\hat{\beta}_{+++})) + \Gamma(rot(\hat{\beta}_{--+})) + \Gamma(rot(\hat{\beta}_{-+-})) + \Gamma(rot(\hat{\beta}_{+--})) - \Gamma(rot(\hat{\beta}_{-++})) - \Gamma(rot(\hat{\beta}_{+-+})) - \Gamma(rot(\hat{\beta}_{++-})) - \Gamma(rot(\hat{\beta}_{---})) = 0.

One easily sees that $\hat{\beta}_{-++} = \hat{\beta}_{+-+} = \hat{\beta}_{++-} = 22\hat{2}121$ and that $\hat{\beta}_{--+} = \hat{\beta}_{+--} = \hat{\beta}_{-+-} = 1\hat{2}12$ is the closure of a periodic braid, as well as $\hat{\beta}_{---} = 1\hat{2}$ (remember that $\Gamma = 0$ for all loops of periodic braids). Hence, (*) could be only satisfied if the integer Laurent polynomial $\Gamma(rot(\hat{\beta}_{+++}))$ is divisible by 3, which is not the case for each of the above polynomials. One can generalize this example to show that the above one-cocycle polynomials are not of finite type of any degree.

Let $n > 3$ be arbitrary. We define a linear combination of configurations $I^1_{(n-2,1)^-} = \sum_i \epsilon_i I^1_i$ in Fig. 4.38.

Theorem 4.6

$$\Gamma^1_{(n-2,1)^-}(S) = \sum_{s_i \in S \cap \Sigma^{(1)}_{tri} of\ type\ (n-2,1)^-} sign(s_i) x^{\sum_i \epsilon_i (\sum_{p_i} w(p_i))}$$

is a one-cocycle of Gauss degree 1, where p_i are the arrows which enter in I_i^1 in the Gauss diagram of s_i.

The proof of the theorem is completely analogous to the proofs of Theorem 4.5 and Proposition 4.10 and is left to the reader.

Again, its mirror image is also a one-cocycle. Moreover, $\Gamma^1_{(n-2,1)-}$ can be easily generalized to polynomial one-cocycles of higher Gauss degrees by introducing appropriate slide moves, along the same lines as Theorem 4.5 and Proposition 4.10.

Example 4.5 *A calculation yields that for $rot(\sigma_3\hat{\sigma_2}\sigma_1^{-1})$ there is a single triple crossing of type $(2,1)^-$ in the loop $\beta\Delta^2\Delta^{-2} \to \Delta^2\hat{\beta}\Delta^{-2}$. It is represented by the closure of the 4-braid $(\sigma_2\sigma_3\sigma_2)\sigma_2^{-1}\sigma_3^{-2}\sigma_1$, where the parentheses as usual refers to the triple crossing before the Reidemeister III move. One easily calculates now that*

$$\Gamma^1_{(2,1)-}(rot(\sigma_1\hat{\sigma_2^{-1}}\sigma_3^{-1})) = -x.$$

Therefore, $\Gamma^1_{(2,1)-}$ is a not always trivial cohomology class for closed pseudo-Anosov 4-braids.

Example 4.6 *Let us consider the reducible 4-braid $\beta = \sigma_3\sigma_2\sigma_3\sigma_1\sigma_2$ which is the 2-cable σ_1 of the 2-braid σ_1. Evidently, the entropy and the simplicial volume of β are trivial, because the JSJ-decomposition of $S^3 \setminus (\hat{\beta} \cup L)$ has only Seifert fibered pieces.*

An easy calculation gives $\Gamma^1_{(n-2,1)-}(rot(\beta)) = x$. Consequently, the vanishing of the entropy and of the simplicial volume of the braid does not imply that the invariant vanishes too.

The 1-cocycle invariants have the following nice property, which can be used to estimate from below the length of conjugacy classes of braids.

Proposition 4.11 *Let the knot $K = \hat{\beta} \hookrightarrow V$ be a closed n-braid and let $c(K)$ be its minimal crossing number, i.e. its minimal word length in B_n. Then all 1-cocycle polynomials of Gauss degree d vanish for*

$$d \geq c(K) + n^2 - n - 2.$$

Proof. Assume that the word length of β is equal to $c(K)$. We can represent $[rot(K)]$ by the following isotopy which uses shorter braids.

$$\beta \to \Delta\Delta^{-1}\beta \to \Delta^{-1}\beta\Delta \to \Delta^{-1}\Delta\beta' \to \beta' \to \Delta\Delta^{-1}\beta' \to \Delta^{-1}\beta'\Delta \to \Delta^{-1}\Delta\beta \to \beta$$

Here, β' is the result of rotating β by π i.e. each $\sigma_i^{\pm 1}$ is replaced by $\sigma_{n-i}^{\pm 1}$. Obviously, $c(\Delta) = \frac{n(n-1)}{2}$. Thus, each Gauss diagram which appears in the isotopy has no more than $c(K) + n^2 - n$ arrows. Indeed, we create a couple of crossings by pushing Δ through β only after having eliminated a couple of crossings before (compare Section 4.7.1). Therefore, for each diagram with a triple crossing there are at most $c(K) + n^2 - n - 3$ other arrows, and hence, each summand in a 1-cocycle of Gauss degree d is already zero if $d \geq c(K) + n^2 - n - 2$. $\square$

(Remember that the degree in Vassiliev's sense of $d(\Gamma(\hat{\beta}))/dx$ evaluated at $x = 1$ is at most $d + 1$, compare Remark 4.4, which leads to the result cited in the Introduction.)

4.8.5   A further refinement of polynomial one-cocycles for closed braids

This subsection contains our strongest polynomial one-cocycles for closed braids.

For simplicity we restrict ourself to the case $n = 3$, but the 1-cocycles can be generalized for $n > 3$ straight forwardly by an interested reader. The main observations for the refinement is that we can consider different weights (or configurations) simultaneously! Moreover, we can refine the weights by taking into account the local types of the triple crossing too.

We consider only triple crossings of the global type $(2,2)^+$ and the corresponding weights which are defined in Fig. 4.39. (Remember that e.g. W_1 means, that we take the sum of the signs of all crossings of marking 1 which are in position relative to the triangle as shown in Fig. 4.39.)

Definition 4.32 *Let γ be a generic oriented loop in M_3^+. The 1-cochain $R(\gamma)$ with values in $\mathbb{Z}[x_1^{\pm 1}, x_2^{\pm 1}, x_3^{\pm 1}, x_4^{\pm 1}]$ is defined by*

$$R(\gamma) = \sum_{p \in \gamma \cap (2,2)^+} sign(p) x_1^{2W_1(p)+w(d)} x_2^{2W_2(p)+w(d)} x_3^{2W_3(p)+w(ml)+w(hm)} x_4^{W_4(p)}.$$

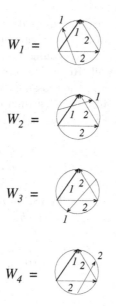

$W_1 =$

$W_2 =$

$W_3 =$

$W_4 =$

Figure 4.39: The weights for the 1-cocycle R

Notice that for $x_2 = x_1^{-1}$, $x_3 = x_4 = 1$ R coincides with our previous 1-cocycle $\Gamma^1_{(2,2)+}$ (compare Fig. 4.36) up to a normalization of the exponents.

Let β be a braid which closes to a knot in the solid torus V. The link $\hat{\beta} \cup$ (core of complementary solid torus) is not invertible in S^3 if and only if β is not conjugate to $\beta_{inverse}$, i.e. the braid which is obtained from reading the braid backwards (compare e.g. [16]).

Theorem 4.7 R *is a 1-cocycle and* $[R] \in H^1(M_3^+; \mathbb{Z}[x_1^{\pm 1}, x_2^{\pm 1}, x_3^{\pm 1}, x_4^{\pm 1}])$ *is in general no-trivial. Moreover,*

$$R(rot(\hat{\beta}))(x_1, x_2, x_3, x_4) = -R(rot(\hat{\beta}_{inverse}))(x_2, x_1, x_3, x_4).$$

Proof. As usual, the weights are chosen such that no arrow can change its position with respect to the triangle (because it would create loops in the diagram which are homologous to 0) besides by exchange moves from passing $\sum^{(2)}_{trans-self}$. In an exchange move the crossing will no longer contribute to the weight W_1 respectively W_2, but the sign of the crossing d has changed too. It follows immediately that $2W_1(p) + w(d)$ and $W_2(p) + w(d)$ stay both invariant. In the case of W_3 the crossing with marking 2 could become the crossing ml or the crossing hm in an exchange move and this time $W_3(p) + w(ml) + w(hm)$ stays invariant. The configuration W_4 is rigid. Reidemeister II moves do not contribute at all. Replacing β by $\beta_{inverse}$ changes the sign for each R III move in the loop and the crossings hm and ml are interchanged (compare Chapter 2). This interchanges then exactly the weights W_1 and W_2. Everything else is preserved and we obtain the formula in the theorem. This completes the proof. $\square$

Example 4.7 *We consider the knot* 8_{17} *as the closure of the 3-braid* $\beta = 2\bar{2}1\bar{2}1\bar{2}1\bar{1}$. *We know of course that* β *is not conjugate to* $\beta_{inverse}$, *because* 8_{17} *is not an invertible knot in* S^3. *We will prove that* $\beta\hat{\Delta}^2$ *is not conjugate to* $(\beta\hat{\Delta}^2)_{inverse}$ *by using* $R(rot(\beta\hat{\Delta}^2))$, *which implies of course that* $\hat{\beta}$ *is not conjugate to* $\hat{\beta}_{inverse}$ *neither.*

We represent the loop $rot(\beta\hat{\Delta}^2)$ *by pushing* Δ^2 *through* $\hat{\beta}$.

Below we give only the sequence of R III moves, which we put into parentheses, and we number them.

$2\bar{2}1\bar{2}1\bar{2}1\bar{1}121121 \rightarrow 2\bar{2}1\bar{2}1\bar{2}1\bar{1}(\bar{1}21)121 \rightarrow_1 2\bar{2}1\bar{2}1\bar{2}1121(\bar{2}12)1 \rightarrow_2$
$2\bar{2}1\bar{2}121(\bar{1}21)121\bar{1} \rightarrow_3 2\bar{2}1\bar{2}12121(\bar{2}12)1\bar{1} \rightarrow_4 2\bar{2}1\bar{2}1(212)1121\bar{1}\bar{1} \rightarrow_5$
$2\bar{2}1\bar{2}1\bar{1}1211(121)\bar{1}\bar{1} \rightarrow_6 2\bar{2}121(\bar{1}21)1212\bar{1}\bar{1} \rightarrow_7 2\bar{2}12121(\bar{2}12)12\bar{1}\bar{1} \rightarrow_8$
$2\bar{2}\bar{1}(212)1121\bar{1}2\bar{1}\bar{1} \rightarrow_9 2\bar{2}\bar{1}1211(121)\bar{1}2\bar{1}\bar{1} \rightarrow_{10} 221(\bar{1}21)1212\bar{1}2\bar{1}\bar{1} \rightarrow_{11}$
$22121(\bar{2}12)12\bar{1}2\bar{1}\bar{1} \rightarrow_{12} 2(212)1121\bar{1}2\bar{1}2\bar{1}\bar{1} \rightarrow_{13} 21211(121)\bar{1}2\bar{1}2\bar{1}\bar{1} \rightarrow_{14}$
$(212)11212\bar{1}2\bar{1}2\bar{1}\bar{1} \rightarrow_{15} 1211(121)2\bar{1}2\bar{1}2\bar{1}\bar{1} \rightarrow_{16} 12112122\bar{1}2\bar{1}2\bar{1}\bar{1}$

We push then Δ^2 *back into the initial position by an isotopy of diagrams in the solid torus.*

One easily sees now that only the moves $3, 4, 7, 8, 11, 12, 15, 16$ *are of the global type* $(2,2)^+$ *(the eight remaining moves are of course of the type* $(1,1)^-$). *We give the corresponding Gauss diagrams of the moves in Fig. 4.40 up to Fig. 4.47.*

Their contributions are the following, where we write the sign in front of the move

$-move\ 3\!:\ -x_1^3 x_2^3$
$+move\ 4\!:\ +x_1^3 x_2 x_3^2 x_4$
$-move\ 7\!:\ -x_1^5 x_2^{-1} x_3^2 x_4$
$+move\ 8\!:\ +x_1^{-1} x_2^3 x_3^4 x_4^2$

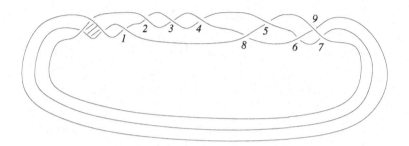

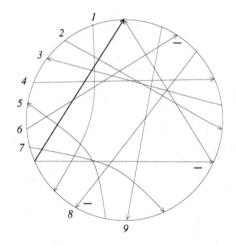

Figure 4.40: The R III move 3

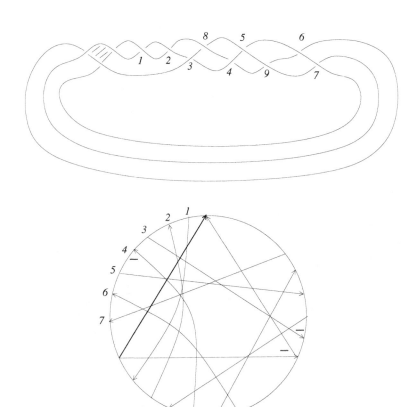

Figure 4.41: The R III move 4

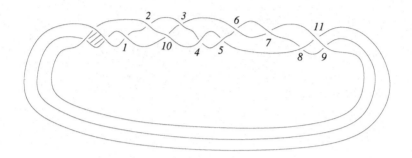

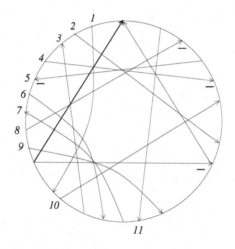

Figure 4.42: The R III move 7

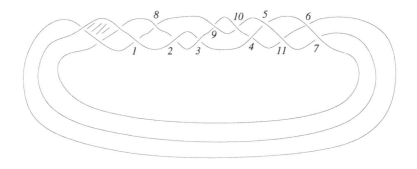

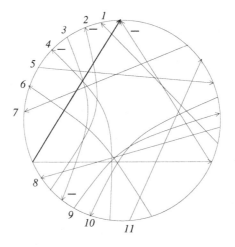

Figure 4.43: The R III move 8

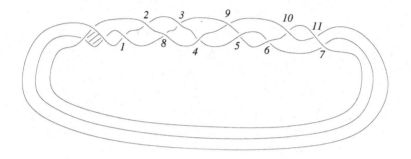

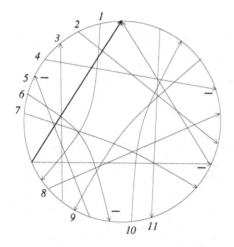

Figure 4.44: The R III move 11

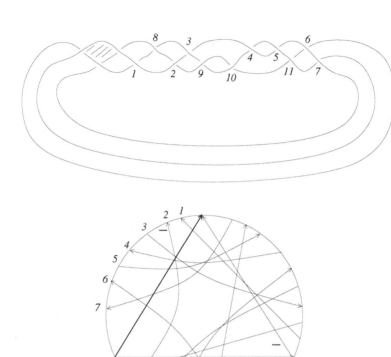

Figure 4.45: The R III move 12

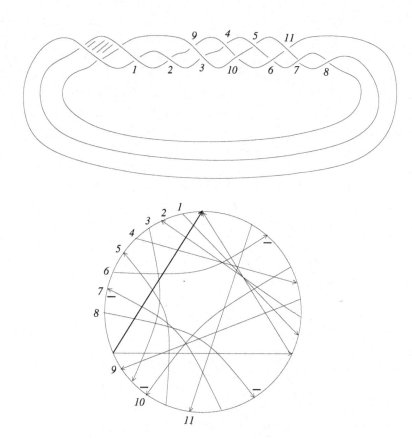

Figure 4.46: The R III move 15

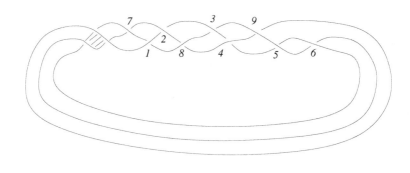

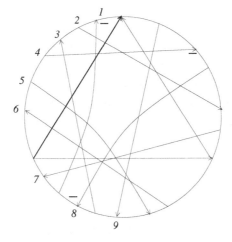

Figure 4.47: The R III move 16

$-move\ 11:\ -x_1x_2x_3^4x_4^2$

$+move\ 12:\ +x_1x_2^5$

$-move\ 15:\ -x_1x_2^3x_3^2x_4$

$+move\ 16:\ +x_1^3x_2x_3^2x_4$

(Notice that there are four positive and four negative moves, as it should be.)

It follows that

$$R(rot(\beta\hat{\Delta}^2))(x_1, x_2, x_3, x_4) = - x_1^5x_2^{-1}x_3^2x_4 - x_1^3x_2^3 + 2x_1^3x_2x_3^2x_4 + x_1x_2^5$$
$$- x_1x_2^3x_3^2x_4 - x_1x_2x_3^4x_4^2 + x_1^{-1}x_2^3x_3^4x_4^2$$
$$= P(x_1, x_2, x_3, x_4) \neq -P(x_2, x_1, x_3, x_4)$$

Consequently, we have shown that $\beta\hat{\Delta}^2$ is not invertible in the solid torus by a polynomial invariant which can be calculated with quadratic complexity.

Notice that the usual quantum knot invariants, as e.g. the skein polynomial in the solid torus (see [48], [52]), can *not* distinguish $\beta\hat{\Delta}^2$ from $(\beta\hat{\Delta}^2)_{inverse}$. This comes from the fact that exactly the same skein calculation applies to both, but for $(\beta\hat{\Delta}^2)_{inverse}$ it ends up with the inverses of the generators of the skein module. The skein module is generated by the closure of permutation braids. But each permutation braid is conjugate to its inverse (i.e. reading it backwards), hence their corresponding closures are isotopic in the solid torus.

Our invariant, the Laurent polynomial $R(rot(\beta\hat{\Delta}^2))(x_1, x_2, x_3, x_4)$, can be calculated with quadratic complexity with respect to the (geometric) braid length of β. In Section 4.9.3 we will distinguish $\hat{\beta}$ from $\hat{\beta}_{inverse}$ with another invariant which can be calculated even with linear complexity!

4.8.6   The finite type invariants from 1-cocycles are not functorial under cabling

Bar-Natan, Thang Le, Dylan Thurston [3] and independently S. Willerton [58] have given a formula for the Kontsevich integral of the cable of a knot in $\mathbb{R}^3$. This formula has the following corollary:

Corollary 4.1 *(Bar-Natan, Thang Le, D. Thurston-Willerton) Let $K, K' \hookrightarrow \mathbb{R}^3$ be knots which have the same Vassiliev invariants up to a fixed degree d. Then, all (the same) cables of K and K' have the same Vasiliev invariants up to degree d.*

In other words, it is useless to cable knots in purpose to distinguish them by Vassiliev invariants.

Our 1-cocycle invariants of degree d form a subset of all finite type invariants of degree d for knots in the solid torus. For example, as already mentioned, there are no non-trivial 1-cocycle invariants at all for closed 2-braids and there are no non-trivial 1-cocycle invariants of degree 1 for closed 3-braids. However, it turns out that cabling is a useful operation for the subset of 1-cocycle invariants of Gauss degree d.

Proposition 4.12 *The closed 2-braids $\hat{\sigma}_1$, $\hat{\sigma}_1^{-1}$ can not be distinguished by any 1-cocycle invariants. However, $Cab_2(\hat{\sigma}_1)$ and $Cab_2(\hat{\sigma}_1^{-1})$ can be distinguished by a 1-cocycle invariant of Gauss degree 1.*

Proof. $Cab_2(\hat{\sigma}_1)$ and $Cab_2(\hat{\sigma}_1^{-1})$ can be represented respectively by the 4-braids $\beta = \sigma_3\sigma_2\sigma_1\sigma_2^2$ and $\beta' = \sigma_3^{-1}\sigma_2\sigma_1^{-1}\sigma_2^{-2}$. A calculation by hand shows:

$$\Gamma_{(2,1)^-}([rot(\hat{\beta})]) = \Gamma_{(3,2)^+}([rot(\hat{\beta})]) = +1$$

and

$$\Gamma_{(2,1)^-}([rot(\hat{\beta}')]) = \Gamma_{(3,2)^+}([rot(\hat{\beta}')]) = -3. \qquad \square$$

Remark 4.8 *Obviously, closed 2-braids are classified by the unique finite type invariant W_1 of degree 1. The braid $\hat{\sigma}_1$ is obtained from $\hat{\sigma}_1^{-1}$ by multiplying σ_1^{-1} with σ_1^2. From this, one easily concludes that for all $k \in \mathbb{Z}$,*

$$\Gamma_{(2,1)^-}([rot(Cab_2(\sigma_1^{2k+1}))]) = \Gamma_{(3,2)^+}([rot(Cab_2(\sigma_1^{2k+1}))]) = 1 + 4k.$$

Therefore, the unique non-trivial 1-cocycle of Gauss degree 1 for the 2-cable of 2-braids classifies also closed 2-braids.

It is an interesting question, whether all finite type invariants of closed braids arise as linear combinations of 1-cocycle invariants of appropriate cables?

4.8.7 Homological estimates for the number of braid relations in one-parameter families of closed braids

Our 1-cocycles can be used in order to obtain information about one-parameter families of closed braids which are knots. Let K be a closed n-braid which is a knot. Let S be a generic loop in $M(K)$.

Definition 4.33 *The *-length $b([S])$ of $[S] \in H_1(M(K); \mathbb{Z})$ is the minimal number of triple points in S among all unions of oriented generic loops S in $M(K)$ which represent $[S]$.*

Proposition 4.13 *Let $a, b \in \{0, 1, \ldots, n\}$ with $a < b$. Then*

$$b([S]) \geq 2 \sum_{(a,b)^+} |\Gamma_{(a,b)^+}([S])| + 2 \sum_{(a,b)^-} |\Gamma_{(a,b)^-}([S])|.$$

Proof. Each $\Gamma_{(a,b)^+}$, $\Gamma_{(a,b)^-}$ is a (in general non-trivial) 1-cocycle of degree 1 (see Proposition 4.8). Each triple point in S contributes by ± 1 to the value of such a cocycle. The inequality follows now from the relations (*) in Proposition 4.8. □

Example 4.8 *Let $\beta = \sigma_1 \sigma_2^{-1} \sigma_3^{-1} \in B_4$. For all $m \in \mathbb{Z}$, we have $b(m[rot(\hat{\beta})]) \geq 4|m|$. This follows immediately from Example 4.2.*

Proposition 4.13 does not contain any information in the case of closed 3-braids (because all 1-cocycles of degree 1 are trivial). Therefore, we use the 1-cocycle Γ of degree 2 for $x = 1$ from Example 4.3. Let $\beta = \sigma_1 \sigma_2^{-1} \in B_3$. One has $\Gamma(rot(\hat{\beta})) = +2$. It is easily seen that this implies $\Gamma!(rot(\hat{\beta})) = +2$ too, where $\Gamma!$ is the mirror image of Γ. Thus, for all $m \in \mathbb{Z} \setminus 0$, $m(rot(\hat{\beta}))$ intersects both types of strata in $\Sigma^{(1)}(tri)$ (compare Fig. 4.24). But the closure of the union of all strata of the same type in $\Sigma^{(1)}(tri)$ are trivial cycles of codimension 1 in $M(\hat{\beta})$ and hence, $m(rot(\hat{\beta}))$ intersects each of these two strata in at least two points. It follows that $b(m[rot(\hat{\beta})]) \geq 4$. The canonical loop shows that this inequality is sharp for $m = \pm 1$.

4.9   Character invariants

In this section we refine our invariants to another class of easily calculable isotopy invariants for closed braids. Let $\hat{\beta}_s, s \in [0, 1]$ be a generic isotopy of closed braids, l an integer, $l.rot(\hat{\beta}_s)$ l-times the canonical loop and $T\tilde{L}(l.rot(\hat{\beta}_s))$ the union of the trace circles of the corresponding resolution of the trace graphs (compare Section 4.1). It follows from Observation 1.1 (compare Remark 4.2) that the trace circles for different parameter s are in a natural one-to-one correspondence. Consequently, we can give *names* x_i to

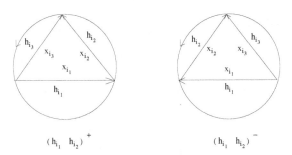

Figure 4.48: Triangles with names of crossings

the circles of $T\tilde{L}(l.rot(\hat{\beta}_0))$ and extend these names in a unique way on the whole family of trace circles.

Let $\{x_1, x_2, \dots\}$ be the set of named trace circles. Obviously, for each circle x_i there is a well defined homological marking $h_i \in H_1(V)$. Let $[x_i] \in H_1(T^2)$ be the homology class represented by the trace circle x_i itself (with its natural orientation induced from the orientation of the trace graph).

4.9.1 Character invariants of Gauss degree 0

In this section, we use the named cycles, i.e. the trace circles, in order to refine the 1-cocycles of Gauss degree 0 which were defined in Section 4.8.3.

Let $X = \{x_1, \dots, x_m\}$ be the set of named trace circles. Let $x_{i_1}, x_{i_2}, x_{i_3} \in X$ be fixed. We do not assume that they are necessarily different. Let $h_{i_1}, h_{i_2}, h_{i_3}$ be the corresponding homological markings.

Definition 4.34 *A character of Gauss degree 0 of $l.rot(\hat{\beta})$, denoted by*

$$C_{(h_{i_1}, h_{i_2}) \pm (x_{i_1}, x_{i_2}, x_{i_3})}(l.rot(\hat{\beta}))$$

or sometimes shortly $C(\hat{\beta})$, is the algebraic intersection number of $l.rot(\hat{\beta})$ with the strata $(h_{i_1}, h_{i_2}) \pm$ in $\Sigma_{tri}^{(1)}$ and such that the crossings of the triple point belong to the named trace circles as shown in Fig. 4.48. We call the unordered set $\{x_{i_1}, x_{i_2}, x_{i_3}\}$ the support of the character $C(\hat{\beta})$.

Remark 4.9 *Evidently, in order to obtain a non-trivial intersection number we need that $h_{i_3} = h_{i_1} + h_{i_2} - n$ in $(h_{i_1}, h_{i_2})^+$ and that respectively $h_{i_3} = h_{i_1} + h_{i_2}$ in $(h_{i_1}, h_{i_2})^-$ (compare Fig. 4.24).*

Notice that for characters of Gauss degree 0 the relations (*) from Proposition 4.8 are no longer valid. For example, $C_{(h_{i_1}, h_{i_2}) + (x_{i_1}, x_{i_2}, x_{i_3})}(l.rot(\hat{\beta}))$ can be non-trivial for $h_{i_1} = h_{i_2}$ and even for $x_{i_1} = x_{i_2}$.

Theorem 4.8 *Let $\hat{\beta}_0$ and $\hat{\beta}_1$ be isotopic closed braids and let $\{x_1, \dots, x_m\}$, $\{x'_1, \dots, x'_{m'}\}$ be the corresponding sets of named trace circles of $T\tilde{L}(l.rot(\hat{\beta}_0))$ respectively $T\tilde{L}(l.rot(\hat{\beta}_1))$.*

Then $m = m'$ and there is a bijection $\sigma : \{x_1, \dots, x_m\} \to \{x'_1, \dots, x'_{m'}\}$ which preserves the homological markings h_i as well as the homology classes $[x_i]$ and such that

$$C_{(h_i, h_j) \pm (x_{i_1}, x_{i_2}, x_{i_3})}(l.rot(\hat{\beta}_0)) = C_{(h_i, h_j) \pm (\sigma(x_{i_1}), \sigma(x_{i_2}), \sigma(x_{i_3}))}(l.rot(\hat{\beta}_1))$$

for all triples $(x_{i_1}, x_{i_2}, x_{i_3})$.

Proof. This is an immediate consequence of Lemma 4.22 and the fact that the trace circles are isotopy invariants of $\hat{\beta}$, see Observation 1.1. $\square$

4.9.2   Character invariants of higher Gauss degree

We refine the results of the Sections 4.8.4 and 4.9.1 in a straightforward way.

Let I^d be one of the configuration of Gauss degree d, which were considered in Section 4.8.4 and let $(x_{i_1}, \dots, x_{i_{d+3}})$ be a fixed $(d+3)$-tuple of elements in X (not necessarily distinct). A *named configuration* $I_{(x_{i_1}, \dots, x_{i_{d+3}}), \phi}$ is the configuration I^d together with a given bijection ϕ of $(x_{i_1}, \dots, x_{i_{d+3}})$ with the $d + 3$ arrows in I^d and such that $x_{i_1}, x_{i_2}, x_{i_3}$ are the arrows of the triangle exactly as in the previous section. The wandering arrow in $I^d_{(2/3n, 2/3n)+, l}$ has the same name in all configurations and corresponding arrows in the bunches of alternating arrows have of course also the same name.

We call the corresponding character polynomial $C^d_{(x_{i_1}, \dots, x_{i_{d+3}})}(l.rot(\hat{\beta}))$.

Different bijections give in general of course different named configurations.

Theorem 4.9 *Let $\hat{\beta}_0$ and $\hat{\beta}_1$ be isotopic closed braids and let $\{x_1, \dots, x_m\}$, $\{x'_1, \dots, x'_{m'}\}$ be the corresponding sets of named trace circles of $T\tilde{L}(l.rot(\hat{\beta}_0))$ respectively $T\tilde{L}(l.rot(\hat{\beta}_1))$.*

Then $m = m'$ and there is a bijection $\sigma : \{x_1, \ldots, x_m\} \to \{x_1', \ldots, x_{m'}'\}$ which preserves the homological markings h_i as well as the homology classes $[x_i]$ and such that for all triples $(x_{i_1}, x_{i_2}, x_{i_3})$

$$C^d_{(x_{i_1}, \ldots, x_{i_{d+3}})}(l.rot(\hat{\beta}_0)) = C^d_{(\sigma(x_{i_1}), \ldots, \sigma(x_{i_{d+3}}))}(l.rot(\hat{\beta}_1))$$

for the configurations I^d which were introduced in Section 4.2.4, and which determine one-cocycles Γ^d.

Proof. This is completely analogous to the proofs of Theorems 4.5, 4.6 and 4.8. $\square$

Let h_i, h_j and a type of triple point, e.g. $(h_i, h_j)^+$, be fixed. It follows immediately from the definitions that

$$(**) \ \Gamma_{(h_i, h_j)^+} = \sum_{(x_k, x_l, x_m)} C_{(h_i, h_j)^+}(x_k, x_l, x_m)$$

where $h(x_k) = h_i$ and $h(x_l) = h_j$.

Hence, character invariants define splittings of one-cocycle invariants. However, the set of character invariants on the right hand side of $(**)$ is not an ordered set.

It follows from Lemma 4.20 that the names x_i are determined by their homological markings h_i, and that there are exactly $n - 1$ trace circles in the case $l = 1$.

However, this is in general no longer true in the case of multiples of the canonical loop.

Let $l \in \mathbb{Z}$ be fixed and let $l.rot(\hat{\beta})$ be the loop which is defined by going l times along the canonical loop. Let $TL(l.rot(\hat{\beta}))$ denote the corresponding trace graph (the t-coordinate in the thickened torus covers now l times the t-circle).

We will show in a simple example that the corresponding monodromy of the crossings (and hence of the trace circles) is in general non-trivial already for $l = 2$.

Example 4.9 *Let $\beta = \sigma_2\sigma_1^{-1}\sigma_2\sigma_1^{-1} \in B_3$. We will write shortly $\beta = 2\bar{1}2\bar{1}$.*

The combinatorial canonical loop for $l = 2$ is given by the following sequence (where we write the names of the crossings just below the crossings).

$2\bar{1}2\bar{1}$ $\rightarrow$ $\bar{1}2\bar{1}2\bar{1}2(\bar{1}1)21$ $\rightarrow$ $\bar{1}2\bar{1}2\bar{1}221$ $\rightarrow$ $\bar{1}2\bar{1}2\bar{1}21(\bar{1}21)$ $*_1 \rightarrow$
$abcd$ $\qquad zyxabcdxyz \qquad zyxabcyz \qquad zyxabcu_1u_1yz$

$\bar{1}2\bar{1}2\bar{1}(212)1\bar{2}$ $*_2 \rightarrow \bar{1}2\bar{1}2(\bar{1}1)211\bar{2} \rightarrow \bar{1}2\bar{1}2211\bar{2} \rightarrow \bar{1}2\bar{1}21(\bar{1}21)1\bar{2}$ $*_3 \rightarrow$
$zyxabcu_1zyu_1 \qquad zyxabzu_1cyu_1 \quad zyxau_1cyu_1 \quad zyxau_2u_2u_1cyu_1$

$\bar{1}2\bar{1}(212)1\bar{2}1\bar{2}$ $*_4 \rightarrow (\bar{1}2\bar{1}121)1\bar{2}1\bar{2}$ $\rightarrow$ $1\bar{2}1\bar{2}$ $\rightarrow$ $\bar{1}2\bar{1}1\bar{2}1(\bar{2}12)1$ $*_5 \rightarrow$
$zyxau_2cu_1u_2yu_1 \qquad zyxcu_2au_1u_2yu_1 \qquad u_1u_2yu_1 \qquad z_1y_1x_1u_1u_2yu_1x_1y_1z_1$

$\bar{1}2\bar{1}1\bar{2}112(\bar{1}1)$ $\rightarrow$ $\bar{1}2\bar{1}1\bar{2}112$ $\rightarrow$ $\bar{1}2\bar{1}1\bar{2}1(121)1$ $*_6 \rightarrow$ $\bar{1}2\bar{1}1(\bar{2}12)12\bar{1}$
$z_1y_1x_1u_1u_2yy_1x_1u_1z_1 \quad z_1y_1x_1u_1u_2yy_1x_1 \quad z_1y_1x_1u_1u_2yy_1x_1v_1v_1 \quad z_1y_1x_1u_1u_2yv_1x_1y_1v_1$

$*_7 \rightarrow \bar{1}2\bar{1}112(\bar{1}1)2\bar{1} \rightarrow \bar{1}2\bar{1}11122\bar{1} \rightarrow \bar{1}2\bar{1}1(121)\bar{1}2\bar{1}$ $*_8 \rightarrow (\bar{1}2\bar{1}121)2\bar{1}2\bar{1}$
$z_1y_1x_1u_1v_1yu_2x_1y_1v_1 \quad z_1y_1x_1u_1v_1yy_1v_1 \quad z_1y_1x_1u_1v_1yv_2v_2y_1v_1 \quad z_1y_1x_1u_1v_2yv_1v_2y_1v_1$

$\rightarrow 2\bar{1}2\bar{1}$
$v_1v_2y_1v_1$

We have the identifications: $d = x$, $b = z$, $x = c$, $y = u_2$, $z = a$, $u_1 = z_1$, $u_2 = x_1$, $x_1 = u_1$, $y_1 = v_2$, $z_1 = y$. *This gives us:*

$2\bar{1}2\bar{1} \rightarrow 2\bar{1}2\bar{1}$
$aacc \qquad v_1v_2v_2v_1$

together with the names $u_1 = z_1 = x_1 = u_2 = y$. *The second rotation gives us:*

$2\bar{1}2\bar{1} \rightarrow 2\bar{1}2\bar{1} \rightarrow 2\bar{1}2\bar{1}$
$aacc \quad v_1v_2v_2v_1 \quad v_3v_4v_4v_3$

together with the identification $v_1 = v_2$, $y_3 = v_4$, $z_2 = v_2$ *and with the names* $u_1 = z_1 = x_1 = u_2 = y$, $u_3 = z_3 = x_3 = u_4 = y_2$. *The monodromy (i.e. how the set of crossings is mapped to itself after the rotation) implies now:* $a = v_3 = v_4 = c$.

Therefore, we have exactly four named cycles: a, v_1, u_1, u_3 *for*

$2\bar{1}2\bar{1} \rightarrow 2\bar{1}2\bar{1}$ *with* $l = 2$.

Consequently, we have

$2\bar{1}2\bar{1} \rightarrow 2\bar{1}2\bar{1} \rightarrow 2\bar{1}2\bar{1}$ *with the names* $aaaa \rightarrow v_1v_1v_1v_1 \rightarrow aaaa$.

Hence, $rot(\hat{\beta})$ *acts by interchanging* a *and* v_1 *(as well as* u_1 *and* u_3). *Consequently, the monodromy is non-trivial in this example.*

Figure 4.49: The names in the first example

4.9.3 Examples for character invariants of trivial Gauss degree

In $l.rot(\hat{\beta})$ each crossing gives rise to exactly $2l(n-2)$ triple crossings (or Reidemeister III moves). Consequently, the calculation of $C_{(h_i,h_j)^{\pm}(x_{i_1},x_{i_2},x_{i_3})}(l.rot(\hat{\beta}))$ is of linear complexity with respect to the braid length of β for fixed l and n. Evidently, $n < c+2$ for a braid which closes to a knot, and hence the invariant is of quadratic complexity with respect to the braid length c for fixed l if n is not fixed.

The following examples for character invariants of Gauss degree 0 are calculated by Alexander Stoimenow using his program in c++. His program is available by request (see [50]).

Let $\beta = \bar{1}2\bar{1}^32^3 \in B_3$. ($\hat{\beta}$ represents the knot 8_9 in the Rolfsen Table.)

We want to show that the link $\hat{\beta} \cup$ (core of complementary solid torus) is not invertible in S^3. This is equivalent to show that β is not conjugate to $\beta_{inverse} = 2^3\bar{1}^32\bar{1}$, i.e. reading the braid backwards (compare e.g. [16]).

Because β is a 3-braid, the homological markings are in $\{1,2\}$.

Character invariants of degree one for $l=1$ and $l=2$ do not distinguish $\hat{\beta}$ from $\hat{\beta}_{inverse}$. However, for $l=3$ we obtain three different named cycles x_1, x_2, x_3 of homological marking 1 and three different named cycles y_1, y_2, y_3 of marking 2.

We consider the set of nine character invariants of Gauss degree 0 which are of the form which is shown in Fig. 4.49, where $i, j \in \{1, 2, 3\}$.

For $\hat{\beta}$ we obtain the set $\{-1, -1, -1, -1, -1, -1, 2, 2, 2\}$ and for $\hat{\beta}_{inverse}$ we obtain the set $\{1, 1, 1, 1, 1, 1, -2, -2, -2\}$. Evidently, there is no bijection of the trace circles for $\hat{\beta}$ and those for $\hat{\beta}_{inverse}$ which identifies the above sets. Consequently, $\hat{\beta}$ and $\hat{\beta}_{inverse}$ are not isotopic in the solid torus.

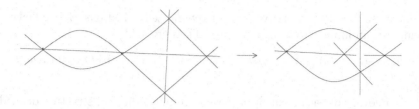

Figure 4.50: A tetrahedron move which creates a generalized trihedron

The knot 9_5 can be represented as a 8-braid with 33 crossings. Character invariants of Gauss degree 0 for $l = 2$ show that the braid is not invertible in the same way as in the previous example.

The knot 8_6 can be represented as a 5-braid with 14 crossings. Character invariants of Gauss degree 0 for $l = 2, 4, 6$ show that it is not invertible as a 5-braid. (Surprisingly, it does not work for for $l = 1, 3, 5$.)

The knot 8_{17} is not invertible as a 3-braid, which is shown with $l = 4$. (It does not work with $l = 1, 2, 3$.)

Let $b \in P_5$ be Bigelow's braid (see [4]). It has trivial Burau representation. Let $s = \sigma_1\sigma_2\sigma_3\sigma_4$. The braids s and bs have the same Burau representation. This is still true for their 2-cables, i.e. we replace each strand by two parallel strands. Character invariants of Gauss degree 0 for $l = 2$ show that the (once positively half-twisted) 2-cables of the above braids are not conjugate, and, consequently, the braids s and bs, which have the same Burau representation, are not conjugate either. *This shows in particular, that even our character invariants of Gauss degree 0 can not be extracted from the Burau representation.*

We will construct a further refinement of character invariants of trivial Gauss degree. The number of triple points in a trace graph can only change by trihedron moves, as follows from Theorem 4.4.

Definition 4.35 *A* generalized trihedron *is a trihedron which might have other triple points on the edges.*

Figure 4.50 shows a tetrahedron move which transforms a trihedron into a generalized trihedron. The generalized trihedron has still exactly two vertices.

Evidently, the number of generalized trihedrons does not change under tetrahedron moves. Let E be the set of all triple points in the trace graph

$TL(l.rot(\hat{\beta}))$ which are *not* vertices of generalized trihedrons. The following lemma is an immediate consequence of Theorem 4.4.

Lemma 4.23 *The set E, and, hence $card(E)$, is an isotopy invariant of closed braids $\hat{\beta}$.*

Moreover, for each element of E we have the additional structure defined before: *type, sign, markings, names.*

It follows from Theorem 4.4 and the geometric interpretation of generalized trihedrons in [20] that the two vertices of a generalized trihedron have always different signs. Consequently, character invariants of Gauss degree 0 count just the algebraic number of elements in E which have a given type and given names. But already the geometric number of such elements in E is an invariant as shows Lemma 4.23.

Definition 4.36 *Let $C^{+(-)}_{(h_i,h_j)+(-)}(x_k, x_l, x_m)$ be the number of all positive (respectively negative) triple points in E of given type $(h_i, h_j)^{+(-)}$ and with given names x_k, x_l, x_m. We call these the* positive (respectively negative) character invariants.

The following proposition is now an immediate consequence of Lemma 4.23 and Definition 4.36.

Proposition 4.14 *The positive and the negative character invariants are isotopy invariants of closed braids for each fixed l.*

Example 4.10 *Let us consider $\beta = \sigma_2\sigma_1^{-1} \in B_3$. Its trace graph $TL(rot(\hat{\beta}))$ is shown in Fig. 4.51. One easily sees that it does not contain any generalized trihedrons. Consequently, all four triple points are in E. There are exactly two names x_1 and x_2. They correspond to the homological markings $h_1 = 1$ and $h_2 = 2$.*

One easily calculates that two of the triple points are of type $(1, 1)^-$ and they have different signs. The other two are of type $(2, 2)+$ and they have different signs too. Consequently, all character invariants of Gauss degree 0 are zero.

However, we have $C^+_{(1,1)-}(x_1, x_1, x_2) = C^-_{(1,1)-}(x_1, x_1, x_2) = 1$, and $C^+_{(2,2)+}(x_2, x_2, x_1) = C^-_{(2,2)+}(x_2, x_2, x_1) = 1$.

Consequently, the positive and negative character invariants contain in this example already for $l = 1$ more information than the character invariants of Gauss degree 0.

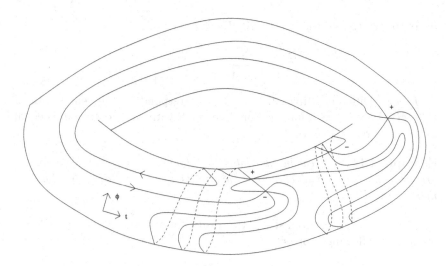

Figure 4.51: The trace graph of $rot(\sigma_2\hat{\sigma}^{-1})$

Unfortunately, there is not yet a computer program available in order to calculate these invariants as well as character invariants of higher Gauss degree in more sophisticated examples.

4.9.4 Homotopical estimates for the number of braid relations in one parameter families of closed braids

In Section 4.8.7 we have used the 1-cocycles in order to estimate the *-length $b([S])$ for classes $[S] \in H_1(M(K); \mathbb{Z})$.

Let $S \subset M(K)$ be a generic loop. We denote its homotopy class for homotopies transverse to $\Sigma_{tan}^{(1)}$ by $[S]_{t-t}$.

Definition 4.37 *The *-length $b([S]_{t-t})$ of a homotopy class transverse to $\Sigma_{tan}^{(1)}$ is the minimal number of triple points in S among all generic loops in $M(K)$ which represent $[S]_{t-t}$.*

Proposition 4.15 *Let $\{x_1, \ldots, x_m\}$ be the set of essential named cycles of S. Let $C_{(x_{i_1}, x_{i_2}, x_{i_3})}(S)$, $\{i_1, i_2, i_3\} \subset \{1, \ldots, m\}$ be the set of all character* *invariants of Gauss degree 0 for S. Then*

$$b([S]_{t-t}) \geq \sum_{\{i_1,i_2,i_3\} \subset \{1,\ldots,m\}} |C_{(x_{i_1},x_{i_2},x_{i_3})}(S)|.$$

Proof. Different triples $(x_{i_1}, x_{i_2}, x_{i_3})$ correspond to different strata of $\Sigma^{(1)}(tri)$. Locally, each intersection index of S with such a stratum is equal to ± 1. The result follows. $\square$

Example 4.11 *One easily calculates that for $S = 2rot(\sigma_2\hat{\sigma_1^{-1}})$ there are exactly eight non-trivial character invariants of Gauss degree 0. Each of them is equal to ± 1.*

This can be generalized straightforwardly for arbitrary l with $|l| \geq 2$. We obtain the following proposition:

Proposition 4.16 *For all integers l such that $|l| \geq 2$, we have:*

$$b([l.rot(\sigma_2\hat{\sigma_1^{-1}})]_{t-t}) = 4|l|.$$

Proof. This follows immediately from Proposition 4.15 together with a direct calculation which shows that

$$b([l.rot(\sigma_2\hat{\sigma_1^{-1}})]_{t-t}) \leq 4|l|.$$ $\square$

Character invariants of higher degree can be used, of course, in the same way as 1-cocycles of higher degree were used to estimate $b([S])$ (compare Section 4.8.7).

4.10   Character invariants of degree one for almost closed braids

Let $K \hookrightarrow V = \mathbb{R}^3 \setminus z - axes$ be an oriented knot such that the restriction of ϕ to K has exactly two critical points. In this case K is called an *almost closed braid*. Necessarily, one of the critical points is a local maximum and the other is a local minimum. In analogy to the case of closed braids we consider almost closed braids up to isotopy through almost closed braids. We consider the (geometric) canonical loop $rot(K)$ and the corresponding

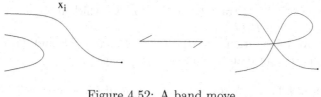

Figure 4.52: A band move

Figure 4.53: An unknot move

trace graph $TL(K)$. $TL(K)$ is an oriented link with triple points and exactly four boundary points (corresponding to the oriented tangencies at the critical points of ϕ).

There are more types of moves for $TL(K)$ as in the case of closed braids (see [20]). But it turns out that only two of these additional moves are relevant for the construction of the invariants.

Definition 4.38 *A band move is shown in Fig. 4.52.*

Definition 4.39 *A unknot move is shown in Fig. 4.53, where the new component is called x_i.*

A band move corresponds to a branch which passes transversely through an ordinary cusp and an unknot move corresponds to the case that the maximum and the minimum of ϕ have the same critical value. Notice, that there are no *extreme pair moves*, i.e. two maxima or two minima of ϕ with the same critical value (for all this compare [20]), because there is only one maximum and one minimum. An extreme pair move induces a *Morse modification of index 1* of the trace graph. Such modifications are not local and they are therefore very difficult to control. We restrict our self to almost closed braids in order to avoid them.

The unknot component x_i of $TL(K)$ which was created by an unknot move, has always $[x_i] = 0$ in $H_1(T^2)$. Notice, that in a band move there is always involved one component x_i which has boundary. Let $X = \{x_1, x_2, \dots\}$

be the set of all closed trace circles of $TL(K)$ which represent non-trivial homology classes in $H_1(T^2)$.

The above observations together with Theorem 4.8 and Theorem 1.10 from [20] imply immediately the following proposition.

Proposition 4.17 *Each character invariant of Gauss degree 0 with names* $x_l, x_k, x_m \in X$ *is an isotopy invariant for almost closed braids.*

Evidently, we can apply Proposition 4.17 to $lrot(K)$ with arbitrary $l \in \mathbb{N}^*$ exactly as in the case of closed braids.

Bibliography

[1] Arnold A., Varchenko A., Gusein-Zade S.: Singularities of Differentiable Maps, Moscow (1982)

[2] Bar-Natan D.: On the Vassiliev knot invariants, Topology 34 (1995) 423-472

[3] Bar-Natan D., Thang T. Q. Le, Thurston D.: Two applications of elementary knot theory to Lie algebras and Vassiliev invariants, Geom. Topol. 7 (2003) 1-31

[4] Bigelow S.: The Burau representation is not faithful for n = 5, Geom. Topol. 3 (1999) 397-404

[5] Birman J.: Braids, Links and Mapping class groups, Annals of Mathematics Studies 82, Princeton University Press (1974)

[6] Birman J., Brendle T.: Braids: A survey, mathGT/0409205 (2004)

[7] Birman J., Gebhardt V., Gonzáles-Meneses J.: Conjugacy in Garside groups III: Periodic braids, J. of Algebra 316 (2007) 746-776

[8] Budney R., Conant J., Scannell K., Sinha D.: New perspectives on self-linking, Advances in Math. 191 (2005) 78-113

[9] Budney R., Cohen F.: On the homology of the space of knots, Geometry & Topology 13 (2009) 99-139

[10] Budney R.: Topology of spaces of knots in dimension 3, Proceedings London Math. Soc. 101 (2010) 477-496

[11] Budney R.: An operad for splicing, J. of Topology 5 (2012) 945-976

[12] Burde G., Zieschang H.: Knots, de Gruyter Studies in Mathematics 5, Berlin (1985)

[13] Carter J.S., Saito M.: Reidemeister moves for surface isotopies and their interpretations as moves to movies, J. Knot Theory Ramif. 2 (1993) 251-284

[14] Damon J.: The Unfolding and Determinacy Theorems for Subgroups of A and K, Mem. Amer. Math. Soc. v.50 (1984)

[15] David J.M.S.: Projection-generic Curves, J. London Math. Soc. (2) v.27 (1983) 552–562

[16] Fiedler T.: Gauss Diagram Invariants for Knots and Links, Mathematics and Its Applications 532, Kluwer Academic Publishers (2001)

[17] Fiedler T.: Isotopy invariants for closed braids and almost closed braids via loops in stratified spaces, arXiv: math. GT/0606443 (48 pp)

[18] Fiedler T.: Quantum one-cocycles for knots, arXiv: 1304.0970v2 (177 pp)

[19] Fiedler T.: One-cocycle invariants for closed braids, arXiv: 1804.03549 (44 pp)

[20] Fiedler T., Kurlin V.: A one-parameter approach to links in a solid torus, J. Math. Soc. Japan 62 (2010) 167-211

[21] Fiedler T., Kurlin V.: Recognizing trace graphs of closed braids, Osaka J. Math. 47 (2010), 885-909

[22] Fox R.: Rolling, Bull.Amer. Math. Soc. 72 (1966) 162-164

[23] Garside F.: The braid group and other groups, Quart. J. Math. Oxf. II Ser. 20 (1969) 235-254

[24] Gordon C., Luecke J.: Knots are determined by their complements. Bull. Amer. Math. Soc. 20 (1989) 83-87

[25] Goryunov V.: Finite order invariants of framed knots in a solid torus and in Arnold's J^+-theory of plane curves, "Geometry and Physics", Lect. Notes in Pure and Appl. Math. (1996) 549-556

[26] Gramain A.: Sur le groupe fondamental de l'espace des noeuds, Ann. Inst. Fourier 27 (1977) 29-44

[27] Goussarov M., Polyak M., Viro O.: Finite type invariants of classical and virtual knots, Topology 39 (2000) 1045-1068

[28] Hatcher A.: A proof of the Smale Conjecture, Ann. of Math. 117 (1983) 553-607

[29] Hatcher A.: Topological moduli spaces of knots, arXiv: math. GT/9909095

[30] Hatcher A., McCullough D.: Finiteness of classifying spaces of relative diffeomorphism groups of 3-manifolds, Geometry & Topology 1 (1997) 91-109

[31] Johannson K.: Homotopy equivalences of 3-manifolds with boundary, Lecture Notes in Math. 761, Springer Berlin (1979)

[32] Jones V.: Hecke algebra representations of braid groups and link polynomials, Ann. of Math. 126 (1987) 335-388

[33] Kashaev R.: The hyperbolic volume of knots from the quantum dilogarithm, Lett. Math. Phys. 39 (1997) 269-275

[34] Kashaev R., Korepanov I., Sergeev S.: Functional tetrahedron equation, Theoret. and Math. Phys. 117 (1998) 1402-1413

[35] Kauffman L.: Knots and Physics (second edition), World Scientific, Singapore (1993)

[36] Kauffman L.: Virtual knot theory, European J. Comb. 20 (1999) 663-690

[37] Ko K.H., Lee J.W.: A Fast Algorithm to the Conjugacy Problem on Generic Braids, math.GT/0611454.

[38] Mancini S., Ruas M.A.S.: Bifurcations of Generic One Parameter Families of Functions on Foliated Manifolds, Math. Scand. 72 (1993) 5-19

[39] Milnor J.: A unique factorization theorem for 3-manifolds, Amer. J. Math. 84 (1962) 1-7

[40] Mortier A.: Combinatorial cohomology of the space of long knots, Alg. Geom. Top. 15 (2015) 3435-3465

[41] Mortier A.: Finite-type 1-cocycles, J. Knot Theory Ramifications 24 (2015) 30 pp.

[42] Mortier A.: A Kontsevich integral of order 1, arXiv: 1810.05747

[43] Morton H.: Infinitely many fibered knots having the same Alexander polynomial, Topology 17 (1978) 101-104

[44] Motegi K.: Knotting trivial knots and resulting knot types, Pacific J. Math. 161 (1993) 371-383

[45] Murakami H., Murakami J.: The colored Jones polynomials and the simplicial volume of a knot, Acta Math. 186 (2001) 85-104

[46] Murasugi K.: On Closed 3-Braids, Memoirs of Amer. Math. Soc., v.151, (1974)

[47] Polyak M., Viro O.: Gauss diagram formulas for Vassiliev invariants, Internat. Math. Res. Notes 11 (1994) 445-453

[48] Przytycki J.: Skein modules of 3-manifolds, Bull. Polish Acad. Sci. Math. 39 (1991) 91-100

[49] Sakai K.: An integral expression of the first non-trivial one-cocycle of the space of long knots in $\mathbb{R}^3$, Pacific J. Math. 250 (2011) 407-419

[50] Stoimenow A.: stoimenov.net/ stoimeno/ homepage

[51] Thurston W.: The Geometry and Topology of Three-Manifolds, http://www.msri.org/ publications/books/gt3m/

[52] Turaev V.: The Conway and Kauffman modules of a solid torus, J. Soviet. Math. 52 (1990) 2799-2805

[53] Turchin V.: Computation of the first non-trivial 1-cocycle in the space of long knots, (Russian) Mat. Zametki 80 (2006), no. 1, 105-114; translation in Math. Notes 80 (2006), no. 1-2, 101-108.

[54] Vassiliev V.: Cohomology of knot spaces // in: Theory of Singularities and its Applications, Advances in Soviet. Math. 1 (1990) 23-69

[55] Vassiliev V.: Combinatorial formulas of cohomology of knot spaces, Moscow Math. Journal 1 (2001) 91-123

[56] Waldhausen F.: On irreducible 3-manifolds which are sufficiently large, Ann. of Math. 87 (1968) 56-88

[57] Wall C.T.C.: Projection Genericity of Space Curves, J. of Topology 1 (2008) 362-390

[58] Willerton S.: The Kontsevich integral and algebraic structures on the space of diagrams, "Knots in Hellas 98", Series on Knots and Everything 24, World Scientific (2000) 530-546

Index

SERIES ON KNOTS AND EVERYTHING

ISSN: 0219-9769

Editor-in-charge: Louis H. Kauffman *(Univ. of Illinois, Chicago)*

The Series on Knots and Everything: is a book series polarized around the theory of knots. Volume 1 in the series is Louis H Kauffman's Knots and Physics.

One purpose of this series is to continue the exploration of many of the themes indicated in Volume 1. These themes reach out beyond knot theory into physics, mathematics, logic, linguistics, philosophy, biology and practical experience. All of these outreaches have relations with knot theory when knot theory is regarded as a pivot or meeting place for apparently separate ideas. Knots act as such a pivotal place. We do not fully understand why this is so. The series represents stages in the exploration of this nexus.

Details of the titles in this series to date give a picture of the enterprise.

Published:

More information on this series can also be found at http://www.worldscientific.com/series/skae

Printed in the United States
By Bookmasters